FOOD CROP MINERAL DEFICIENCY AND DISTURBANCE STRESS MITIGATION IN TEMPERATE CLIMATIC REGIONS BY ECONOMICAL AND ENVIRONMENTAL VALORIZATION OF AGRICULTURAL BY-PRODUCTS

Food Crop Mineral Deficiency and Disturbance Stress Mitigation in Temperate Climatic Regions by Economical and Environmental Valorization of Agricultural By-Products

Edward Someus

Nova Science Publishers, Inc.
New York

For permission to use material from this book please contact us:
Telephone 631-231-7269; Fax 631-231-8175
Web Site: http://www.novapublishers.com

LIBRARY OF CONGRESS CATALOGING-IN-PUBLICATION DATA

Someus, Edward.
Food crop mineral deficiency & disturbance stress mitigation in temperate climatic regions by economical & environmental valorization of agricultural by-products / Edward Someus.
p. cm.
Includes index.
ISBN 978-1-60692-243-9 (softcover)
1. Food crops--Fertilizers--Economic aspects. 2. Food crops--Fertilizers--Environmental aspects. 3. Organic wastes as fertilizer. 4. Agricultural wastes. I. Title. II. Title: Food crop mineral deficiency and disturbance stress mitigation in temperate climatic regions by economical and environmental valorization of agricultural by-products.
SB175.S63 2009
631.8--dc22

2008044289

Published by Nova Science Publishers, Inc. ✝ New York

CONTENTS

PREFACE

The innovative 3R "Recycle-Reuse-Reduce" AGROCARBON technology enables recycling of agricultural organic and mineral by-products which provide carbon products for soil amendment and restoration of soil natural balance. This book explains how the input by-product feed streams, such as refuse grain, sawdust, food grade animal bone meal, food processing and/or other agro by-products, converted and valorized to added value agrocarbon carboniferous products. This innovative technology provides surface-modified charcoals and minerals for plant availability and post- processing of the chars by integrated biotechnological means. The process upgrades by-products to high added-value biological control, plant growth promotion and natural fertilization combined products for environmentally friendly vegetable cultivation, with carbon sequestration potential. The 3R technology is a horizontally arranged and indirectly heated low temperature zero emission carbonization system (operating under vacuum, up to 850 °C±50°C material core temperature) and directly integrated novel agro biotechnological processing units of agrocarbon specific solid state fermentation and formulations. *Performance*: food crop mineral deficiency and disturbance stress mitigation in temperate climatic regions by restoration of soil natural balance have been demonstrated.

Input feed streams: these consisted of low value organic and/or inorganic by-products such as refuse grain, sawdust and/or high phosphorous content animal bone meal, and/or other by products that can be valorization transformed by added-value integrated thermal and biotechnological means.

The industrialized 3R biotechnology integrated biochar production is a modern zero emission solution, a process whereby all and any output products are recycled and reused, aiming for prevention-protection-preservation approaches.

The output products are different types of biotechnology-specific solid carrier composites and adapted microbiological fungus and/or bacteria strain consortia. Depending on the soil and climate application scenario conditions, different types of soil and climate-relevant 3R NPK products can be made.

The application objective of the products are the natural balance and functionality restorations of degraded temperate agriculture soils with controlled microbiological activity and precision farming nutrient supply. Further objectives are the promotion of humus building and mineral mobilization towards plant availability for sustainable, improved, economical and ecological food crop production in the fields of organic and low-input, low-greenhouse gas farming, while carbon sequestration is also targeted.

The application targets combined effects, such as plant growth promotion, biological control against soil-borne plant pathogens and natural NPK fertilization, especially sequenced mobilized phosphorus supply and improved nutrient use efficiency.

The application sectors are organic farming and/or low-input farming for environmentally-friendly vegetable cultivation and other food crop productions.

STATUS: a "product like" field demonstration plant has been developed, successfully tested, with scale-up optimization and comprehensive industrialized engineering design made for a 30,000 m^3/year input feed stream in accordance with modern US/EU industrial norms and standards. This patented original solution is available for licensing and technology transfer.

INTRODUCTION

The innovative 3R AGROCARBON technology and methods are patented and IPR protected original solutions. The application objective is the *mitigation of food crop mineral deficiency and disturbance stress in temperate climatic regions by economical and environmental valorization* of agricultural by-products. The 3R process is performed by *integrated low temperature carbonization and biotechnological recycling of minerals and/or carboniferous* biotechnological carriers from organic and inorganic agricultural by-products. The input feed stream is high phosphorus content animal bone meal, refuse grain, sawdust, food processing and/or other by-products. This innovative technology provides surface-modified biochars and minerals for use as plant-available microbiological carriers. The innovative design is a zero emission industrial process. Natural soil-borne microbiological fungus and/or bacteria aerobic strains are selected for solid carrier entrapment and formulated for ambient temperature long term storage. The process upgrades by-products to high added-value and safe biological crop protection, plant growth promotion and natural NPK fertilization products for environmentally-friendly vegetable cultivation, with carbon sequestration potential. The biological-control effect targets primarily combat crown rot and damping off plant pathogens with preventive measure, enhancement of plant natural resistance and soil natural balance restoration.

The advanced science and engineered design includes; added-value by-product transformation by a specific thermal desorption/low temperature carbonization system, selection of micro-organisms for solid state fermentation and formulation with combined biological-control and NPK mineral-mobilization effects, solid-state-fermentation-formulation, process optimization and scale-up, mineral-carbon soil kinetics evaluation, comprehensive risk assessment, validation/field demonstration under several Europe-wide climatic/soil conditions, authority permission, cost-benefit-analysis, consumer acceptance and EU/USA market evaluation. The final achievements are prevention-protection-preservation approaches, aiming for environmentally-friendly and healthful food production for the low-input and organic farming horticultural production sectors.

"A product like" field demonstration plant has been developed, successfully tested, with scale-up optimization and comprehensive industrialized engineering design made for a 30,000 m^3/year input feed stream modern US/EU industrial norms and standards. This patented original solution is available for licensing and technology transfer.

ABBREVIATIONS AND TERMS

ABC	Animal bone charcoal mineral, which has been produced by a specific 3R carbonization technology surface modification method from animal bone meal, carbon/mineral-based agricultural and food industrial waste and by-products, aiming for agro biotechnological added-value re-use in food production cultivation applications. Different types of ABC are produced and optimized to meet the regional application need for different types of soils under different worldwide climatic conditions, including scenarios for drought under climate change conditions. Scheduled doses are from 400 kg/ha to 1250 kg/ha.
ADE	Amazonian dark earth, also known as terra preta
AMF	Arbuscular mycorrhiza fungi
Anthrosols	The anthrosols are soils that have been formed or profoundly modified through long-term human activities, such as addition of organic materials or household wastes, irrigation or cultivation, where human and time factors have impacted the genesis. It has been identified in tropical/subtropical weathered soils (Brazil, Japan, China), but is also associated with infertile soils such as arenosols, podzols and albeluvisols, as well as with wetland soils (fluvisols, gleysols). Hortic anthrosols occur everywhere where long-term intensive cultivation has taken place [90].
3R AGROCARBON	The agricultural product family of the 3R technology outputs, a mineral/charcoal product containing abundant microorganisms. The specific biochar-charcoal may be of plant and/or animal bone origin.
3R technology	Recycle-Reduce-Reuse, integrated carbonization and agro biotechnology for added "carbon economy" value processing of agricultural and food industrial waste and by-products, done by specific pyrolysis and SSFF technologies. Inventor: Edward Someus, a Swedish environmental engineer. Patented.
3R group	3R Environmental Technologies Ltd, a holding company for rights, knowledge, know-how and engineering design for the 3R technology and 3R-AGROCARBON products. Owner and general manager: Edward Someus. http://www.3ragrocarbon.com
3R-Tn	*Trichoderma harzianum sp.* soil fungus strain selected by the 3R group with

	"n" serial number
Cd	Cadmium
CFU	Microorganism colony-forming units
Chemosynthetic fertilizers	Nitrogen fertilisers, which have been produced by an industrial process and phosphorus fertilisers, which have been produced from mined phosphorus rock by an industrial process involving treatment with strong acid(s).
EU	European Union
ft^3	Cubic feet
EPA	United States Environmental Protection Agency
GHG	Greenhouse gas
GWP	Global warming potential
kBq	Kilo becquerel
MJ	Mega joule
NOx	Oxides of nitrogen
OECD	Organisation for Economic Co-operation and Development
Organic farming	All agricultural farming systems that respects the biological relationship that exists in nature, thereby fostering natural resource and environmental conservation, while promoting the environmentally, socially and economically sound agro productions.
PR	Phosphate rock
P	Phosphorus
3R-AGROCARBON product	Brand name of the 3R technology output product
PROTECTOR project	Applied scientific development project acronym, contracted 2005-2008 with the European Commission, DG RTD E.4, SDME 08/16 Directorate E - Biotechnology, Agriculture, Food Unit 4, Contract No. Food-2005-514082. Full title: Recycling and upgrading of bone meal for environmentally friendly crop protection and nutrition. http://www.terrenum.net/protector, http://www.3ragrocarbon.com
RTD	Applied scientific research and technical development
SSFF	Solid state fermentation and formulation, specifically developed for ABC biotechnological processing, including formulations of individual stains or formulations of strain consortium or co-composting with ABC mineral/charcoal-containing abundant microorganisms
SOM	Soil organic material
SRM	Specified risk material, food safety standard and classification for handling of

	waste materials from food processing
Terra preta	Very dark and fertile anthropogenic soils found in the Amazon Basin with high charcoal content. It is also known as "Amazonian dark earth" (ADE), or "Indian black earth" or "terra preta de índio."
Tg	10^{12} g or million metric tons
WSP	Water-soluble phosphorus
Temperate climatic	The temperate latitudes of the globe lie between the tropics and the polar circles and include the European Union, USA, Canada, Japan, China, Russia. The north temperate zone extends from the Tropic of Cancer (at about 23.5 degrees north latitude) to the Arctic Circle (at approximately 66.6 degrees north latitude). The south temperate zone extends from the Tropic of Capricorn (at approximately 23.5 degrees south latitude,) to the Antarctic Circle (at approximately 66.5 degrees south latitude).
Temperate agriculture soil	Different types of human-influenced and intensively-cultivated soils for food crop and energetic biomass production in temperate climatic regions usual characteristics of these soils include the erosion, degradation after long periods of intensive agricultural cultivations and depleted carbon stocks and nutrient deficiency in comparison with native ecosystems
TOC	Total organic carbon
VOC	Volatile organic compounds

Chapter 1

GENERAL CONSIDERATIONS: THE MULTIDIMENSIONAL CONCERNS OF INTENSIVE CROP PRODUCTION

Agribusiness is industry dependent on manufactured chemical inputs for high yields. For every additive beneficial to productivity, there are by-products requiring extensive environmental services to render them inert. Radioactive gypsum stacks, residual pesticides and increasing nutrient loading to lakes and rivers are examples [10].

In the last 50 years, mankind has modified the environment more than during the whole history of humanity, and the artificial modification process by intensive agriculture production played decisive role. The increased population pressure, reduced length of fallow, deforestation, biodiversity decreases, over-use of chemosynthetic substances and chemosynthetic mineral NPK inorganic fertilizers in combination with improper agricultural practices have led to widespread soil degradation in many parts of the world, whereas the biodiversity loss disrupts the functioning of ecosystems. An important manifestation of this environmental damage is the inadequate replenishment of soil nutrients and organic matter. In particular, P deficiency is becoming critical in many soils. Moreover, because of complementarities in the uptake of plant nutrients, this deficiency threatens to disturb the viability of applying other nutrients. In order to preserve the sustainability of agriculture and safeguard the livelihood of large segments of the rural population, there is an urgent need to rebuild soil fertility and so maintain and improve current levels of productivity and farm income [38].

A doubling in global food demand projected for the next 50 years poses huge challenges for the sustainability both of food production and of terrestrial and aquatic ecosystems and the services they provide to society. New incentives and policies for ensuring the sustainability of agriculture and ecosystem services will be crucial if we are to meet the demands of improving yields without compromising environmental integrity or public health. Modern agriculture now feeds 6 billion people. By 2050, global population is projected to be 50% larger than at present and global grain demand is projected to double [97].

Global food stocks are running low and rich nations should not take security of supplies for granted. There are *food supply uncertainties* we face in the twenty-first century for which there are several reasons:

1) increasing use of crops for bio-energy

2) increasing demand for meat and milk products in the developing world
3) poor harvests around the world following droughts and floods

The global area for food production will decline as farmland is lost to housing, bio-energy cropping and, ultimately, sea level rise as well. However, it will not be acceptable to increase production without regard for the environment, and we will increasingly demand food that is safe and contributes to healthful diets, while it is predicted that an excessive supply of food just exports environmental problems. About half of global usable land is already in pastoral or intensive agriculture [96].

Furthermore, only recently have we started to think about how agriculture should contribute to managing climate change by controlling the release of greenhouse gases and by storing carbon in the soil, while we also need diversity in agriculture and restoration of soil natural balance after decades of high-input farming. There will be no single path to sustainability; organic, low-input farming hi-tech plant and animal breeding will be part of the mix, possibly concentrated in different regions and serving different markets. Recycling, reduction and reuse of waste in the food chain is another important issue that needs to be addressed. There is no point in producing more food from the land without trying to use what we already have more efficiently [26]. Although we can produce enough food to feed today's population, that achievement has come at the cost of an ever-increasing impact on Earth's sustainability. The striking increase in the use of nitrogen (N) and phosphorus (P) fertilizers between 1960 and 2000 by intensive agricultural practices has led to degradation of air and water quality [12] [98]. In addition to causing the loss of natural ecosystems, agriculture adds globally significant and environmentally detrimental amounts of nitrogen and phosphorus to terrestrial ecosystems [110]. Agriculture will require an additional 40 and 20 Tg (10^{12} g or million metric tons) of N and P fertilizer, respectively, applied to agricultural soils to meet food production needs in 2040 [12].

These considerations mean, that we will need to produce more healthful food with higher nutritional content per hectare from the farmland that remains, but based on long-term sustainable environmental and ecological systems.

Table 1. Agriculture production and resource use, the recent past to the near future [107]

Item	1960	2000	2030-2040
Food production (Mt)	1.8 x 109	3.5 x 109	5.5 x 109
Population (billions)	3	6	8 (maybe 10)
Irrigated land (% of arable)	10	18	20
Cultivated land (hectares)	1.3 x 109	1.5 x 109	1.8 x 109
Water-stressed countries	20	28	52
N fertilizer use (Tg)	10	88	120
P fertilizer use (Tg)	9	40	55-60

In this context we define sustainable agriculture as practices that meet current and future societal needs for food and fibre, for ecosystem services, for healthy lives, and that do so by maximizing the net benefit to society when all costs and benefits of the practices are considered [97]. The main environmental impacts of agriculture come from the conversion of natural ecosystems to agriculture, from agricultural nutrients that pollute aquatic terrestrial habitats-groundwater, and from pesticides, especially bio-accumulating or persistent organic agricultural pollutants. Agricultural nutrients enter other ecosystems through leaching, volatilization and the waste streams of livestock and humans. Pesticides can also harm human health, as can pathogens, including antibiotic-resistant pathogens associated with certain animal production practices [97].

Continuous cropping and inadequate replacement of nutrients removed in harvested materials or lost through erosion, leaching or gaseous emissions deplete fertility and cause soil organic matter levels to decline, often to half or less of original levels [63].

The effects of land degradation on productivity can be compensated for by increased fertilization, irrigation, and disease control, which increase production costs [70].

The increased urbanisation of populations creates greater dependence upon mineral fertilizers as a greater proportion of wastes from food distribution and consumption is generated remotely from farms [93]. Soil nutrient management needs to satisfy the multiple agricultural goals of increasing productivity, being economically viable, and sustaining environmentally healthy soil and crop systems. Vegetables belong to the group of intensive humus consumers, taking up to approx. 760 kg Humus-C per ha and year. Regulation of soil states that humus has to be compensated at least every third year to zero losses [13] .Sustainable agriculture is that which is managed toward greater resource efficiency and conservation while maintaining an environment favourable for the evolution of all species.[36].

Chapter 2

TRENDS IN FERTILIZER USE

Until about 1950, farmers throughout the world relied on animal wastes and/or crop rotation with nitrogen-fixing crops to keep the soil fertile. Though these practices are still common in some locations, much of the world's agriculture now uses chemosynthetic substances and chemosynthetic inorganic-mineral NPK fertilizers to supplement and restore soil nutrients. Between 1920 and 1960, phosphate fertilizer consumption was greater than that of either nitrogen or potash fertilizer. After 1960, nitrogen fertilizer consumption quickly surpassed that of phosphate and potash fertilizer. Today, nitrogen fertilizer consumption is about 2.5 times higher than that of phosphate fertilizer, and almost four times higher than that of potash fertilizer. Total fertilizer consumption is about 80 times its 1920 rate [29].

The major plant nutrients (nitrogen, phosphorus, and potassium) are critical for maintaining crop yields but have also been associated with the impairment of numerous streams, lakes, and aquifers. Commercial fertilizer is a major agricultural input; farmers typically spend over 10 billion dollars annually on commercial fertilizer, although fertilizer use and prices vary from year to year. However, fertilizer prices, especially for nitrogen, have been volatile and have risen rapidly in recent years. Despite increased fertilizer prices and growing concern about environmental risks from fertilizer use, the use of nutrient management practices on major crops has changed little since the early 1990s. U.S. commercial fertilizer use peaked in 1981 at over 23 million nutrient tons, but has seldom exceeded 22 million tons since then [114]. Since 1990, the primary fertilizer products purchased by farmers have exhibited much different nominal price patterns. Potash prices, in the form of potassium chloride (KCl), have been stable. A major source of phosphate, diammonium phosphate (DAP) has demonstrated much more price variability—peaking in 1995 at near-record levels, declining steadily through 2002, then rising again in 2003 and 2004. Nitrogen product prices, such as those for anhydrous ammonia, have shown the greatest volatility during the last several years—a spike in 1995, a fall to historically low levels by 1999 and 2000, record-high prices in 2001, followed by a large decline in 2002 and recovery in 2003. During 2001 and 2003, nitrogen fertilizer prices rose dramatically in concert with natural gas prices. In response to higher prices in the United States, nitrogen-based fertilizer imports have increased significantly in recent years [114].

Fertilizer consumption reached its maximum in the early 1980s for developed areas, such as Western Europe and North America, and has since levelled off somewhat. Asia has seen a significant rise in fertilizer consumption, increasing from about 1 million to 38 million metric tons of nutrients between 1960 and 1999. The consumption rise in South Asia is also notable—increasing from less than 1 to over 22 million metric tons of nutrients. Other developing regions, including Latin America, East Asia, the Near East, Africa, and Oceania have increased fertilizer consumption steadily since the middle of the twentieth century [29]. Developing Asia uses an enormous quantity of nitrogen fertilizer; in fact, it consumes more than three times as much nitrogen fertilizer as North America. The nitrogen–phosphate–potash ratio of fertilizer consumption for developing Asia is roughly 6–2–1; for North America, the ratio is roughly 2½–1–1. Latin America consumes the three types of fertilizers in more equal quantities, with nitrogen use only slightly higher than phosphate use, and phosphate use only slightly higher than potash use. Worldwide, the current nitrogen–phosphate–potash ratio of fertilizer consumption is about 8–3–2 [29].

Chapter 3

THE ENERGY DEMAND OF INTENSIVE FARMING

Direct energy consumption in the agricultural sector includes use of gas, diesel, liquid petroleum, natural gas, and electricity. Indirect energy use involves agricultural inputs, such as chemosynthetic nitrogen fertilizers, which have a significant energy component associated with their production. Since 1992, direct fuel and electricity expenses for U.S. farms have averaged around seven percent of total operating costs. Diesel fuel and gasoline are widely used for tillage, planting, transportation, and harvesting. Electricity, gas, and natural gas are primarily used in drying; irrigation; operation of livestock, poultry, and dairy facilities; and on farm processing and storage of perishable commodities. Expenses from indirect energy use increase total energy expenditures to 15 percent of operating costs and the accelerating energy prices in 2008 and onward will further increase the energy consumption expenditures in the agriculture sector. Fertilizers embody the most energy among production inputs because natural gas is the primary input (70-90 percent of the cost of producing nitrogen fertilizer) [85].

Energy consumed in primary agricultural production is classified as either direct or indirect use. Direct energy use is the consumption of fossil fuels for various farm operations such as fuel for tractors or fuel for crop drying. Indirect energy use is the conversion of fossil fuels into other products such as fertilizers or pesticides [65].

Intensive farming is concerned above all with productivity and uses a high level of inputs to achieve it. The inputs are usually in the form of chemicals, fertilisers, pesticides and growth-regulators produced by energy intensive industrial processes and additional energy in the form of high levels of mechanisation. Intensive farming does provide large quantities of relatively cheap food — but often at a cost to the environment, nutrient content or animal welfare.

Indirect energy use involves agricultural inputs, such as chemosynthetic fertilizers, which have a significant energy component associated with their production. Expenses from indirect energy use increase total energy expenditures to 15 percent of operating costs. Changes in energy prices may have a greater effect on producers of major field crops [85].

According to the OECD, the absolute energy consumption per hectare increased in OECD countries by 39% between 1970 and 1989. On average, some 1734 MJ are consumed per hectare of agricultural land, rising to 46,400 MJ for the highest consumer, Japan [71]. Traditional P fertilizer production is based on chemical processing of insoluble mineral phosphate high-grade ore, which includes an energy intensive treatment with sulphuric acid at high temperature [109]. The manufacture, distribution, delivery and application of different fertilizers and pesticides are energy intensive. The fertilizer and pesticide prices have been continually increasing due to increased energy costs for production, especially natural gas, increased transportation costs and increased demand. The higher energy costs resulting higher fertilizer and pesticide costs which are reducing the profitability of the crop and vegetable production.

ENERGY INTENSITY OF CHEMOSYNTHETIC FERTILIZATION

Primary production in the agricultural industry is highly dependant upon fossil fuels. Inorganic chemosynthetic fertilizers are major consumers of energy in the agricultural sector. In the United States, inorganic fertilization accounts for about a third of total energy input to crop production. In contrast to tractors, irrigation pumps, and other types of equipment, fertilizers are indirect energy consumers. That is, the bulk of energy use associated with fertilizers is not consumed directly at the agricultural site, but indirectly during its production, packaging, and transportation to the site. Additional energy is then used on-site during fertilizer application [29].

Most chemosynthetic fertilizer energy use is attributable to the production of nitrogen fertilizers with natural gas. Natural gas is the principal energy resource for creating anhydrous ammonia, a key nitrogen fertilizer. Over 90% of nitrogenous fertilizers contain ammonia and/or other fertilizer elements derived from ammonia (e.g., ammonium nitrate, sodium nitrate, calcium nitrate, ammonium sulphate, ammonium phosphates, and urea). Producing ammonia is a very energy intensive process; it requires about 1090 to 1250 m^3 of natural gas to produce one metric ton of anhydrous ammonia (35,000 to 40,000 ft^3 of natural gas per short ton). In 1999, over 15 million metric tons (17 million short tons) of ammonia were produced in the United States, with almost 90% of that going to the fertilizer industry. This amount corresponded to about 3% of total US natural gas production [29].

Each of the three primary nutrients in inorganic fertilizers has a different set of energy requirements during its life cycle. However, these requirements can be separated into four main stages: production, packaging, transportation, and application. Table 2 summarizes the world average energy requirements by nutrient type and life cycle stage for inorganic fertilizers. The table clearly shows the relatively high energy intensity of nitrogen production. This corresponds to almost 90% of nitrogen's total energy requirement. In contrast, the production of phosphate and potash account for only about 45% of the total energy requirement for these nutrients. Moreover, the energy requirement for nitrogen fertilizer is 4.5 times that of phosphate fertilizer, and 5.7 times that of potash fertilizer [29]. Table 2 summarizes the energy requirements to produce, package, and apply inorganic fertilizers.

ENERGY INTENSIVE PRODUCTION OF MAJOR N FERTILIZERS

Ammonia is a basic industrial material and is the material from which other nitrogen fertilizers are made. It has been manufactured on a commercial scale since 1913, and the process used to produce it, originally developed by Fritz Haber, has changed little since that time. In the Haber process, gaseous hydrogen (H_2) and nitrogen (N_2) are combined under high temperature and pressure conditions to produce ammonia.[22].

Table 2. Energy Requirements to produce, package, transport, and apply inorganic chemosynthetic fertilizers. Source: data compiled from [40].

	Energy reqirement (world average) Btu/lb(kj/kg)		
	Nitrogen	Phosphate	Potash
Produce	29,899 (69530)	3,313 (7700)	2,753 (6,400)
Package	1,119 (2,600)	1,119 (2600)	774 (1,800)
Transport	1,936 (4,500)	2,452 (5,700)	1,979 (4,600)
Apply	688 (1,600)	645 (1,500)	430 (1,000)
Total	33,642 (78,230)	7,529 (17,500)	5,936 (13,800)

Ammonia production is an energy intensive process, not only because of its temperature requirements, but also because it requires large quantities of purified H_2 that is usually derived from natural gas, or methane (CH_4). Naphtha and coal are also used as feedstocks in some parts of the world. Hydrogen is produced from methane (CH_4) in a multi-step process. First, a "synthesis gas" is produced by "reforming" methane and steam to hydrogen and carbon monoxide (CO) over a catalyst at 1300-1600° F. This synthesis gas is mixed with air in a second reforming reaction over another catalyst, producing a mixture of N_2, H_2, CO, CO_2 and steam. The carbon monoxide and steam then undergo a final "shift conversion", producing carbon dioxide and more hydrogen. This final synthesis gas is then purified to remove CO_2 that would poison the ammonia-synthesis catalyst. The purified N_2-H_2 synthesis gas is compressed to pressures of 2000-5000 psig (depending on process) and passed over an iron catalyst where ammonia is formed.[22].

The price of chemosynthetic nitrogen fertilizers is directly related to the price of natural gas (methane). Manufacturing one ton of anhydrous ammonia fertilizer requires 33,500 cubic feet of natural gas. This cost represents most of the costs associated with manufacturing anhydrous ammonia. Most of the other popular forms of nitrogen fertilizer are made with anhydrous ammonia. Urea is formulated by a reaction between anhydrous ammonia and carbon dioxide at high temperature and pressure. Ammonium nitrate is formulated by combining anhydrous ammonia and nitric acid in a very corrosive manufacturing climate. Solution liquid fertilizers (28 to 32 percent nitrogen) are composed of one-half urea and one-half ammonium nitrate. It's pretty hard to apply a nitrogen fertilizer formulation that doesn't have natural gas in its manufacturing process [28].

Since natural gas is such a critical resource in fertilizer production, natural gas price fluctuations have a dramatic effect on fertilizer costs. As energy costs continue to rise, and the demand for fertilizers increases, this effect is becoming more pronounced.[29].

Energy conservation in primary agricultural production must make the most efficient use of non-renewable energy resources while integrating on-farm resources such as biological cycles and controls [35]. At the same time, the conservation measures must strike a balance with the industry-wide goal of maximizing food production and economic returns from a given land base.

Chapter 4

NUTRIENT DISTURBANCE IN AGRICULTURAL AND FOREST SOIL

The degradation of forest ecosystems may be attributed to various natural and anthropogenic factors such as climatical extremes, biotic stresses, selection of tree species, harvesting regimes, litter raking, off-site amelioration measures, former land use, air pollutant deposition and soil acidification, as caused by internal and external processes. An important factor for a loss of tree vitality are nutritional disturbances, eventually leading to declining stand stability and productivity [47].

Both salt-like and lime fertilizers contribute positively to revitalization of nutrient-deficient stands. Experience indicates that forest liming may indeed counterbalance the progress of soil acidification as caused by high H deposition [47].

Besides fertilization, forest regeneration with site-adapted tree species may greatly contribute to the rehabilitation of forest ecosystems, since the capacity for storage of C and N and hence, for a closer nutrient cycling, may be improved significantly through this approach [47].

In central Europe, a new type of forest damage has been observed since the mid 1970s. Various investigations indicate that the declines are frequently associated with nutritional disturbances. Good correlations between the site-specific substrate chemistry and the actual nutritional status of the trees/stands were found. To explain the sudden and widespread appearance of the forest declines, adverse anthropogenic impacts mainly due to elevated emissions of air pollutants and their atmospheric derivatives are hypothesized in combination with natural stress factors. Causal mechanisms include soil degradation due to accelerated soil acidification and increased nutrient leaching from the canopy of forest stands. Fertilization and liming experiments have demonstrated that a fast and sustained revitalization and re-stabilization of declining forest ecosystems marked by nutritional disturbances can be achieved [46].

In agriculture, the low uptake and utilization of nutrients out of the added fertilizers results in lower production and appearance of nutrient stress symptoms. The poor efficiency of added fertilizers is mainly due to imbalance fertilization or improper management practices. When the soil is deficient in a particular nutrient element and the crop grown on such soil shows visible deficiency symptoms, there is an urgent need to correct it to improve the crop production. Methods of correcting nutrient deficiencies vary according to agro-climatic regions, the socio-economic situations of the region, the magnitude of disorder, and nutrient element involved.[50].

The tolerance in a given plant species or genotype to nutrient stress is closely related to its nutrient use efficiency. Genotype differences in nutrient use efficiency are linked with root nutrient acquisition capacity, or with utilization by the plant, or both. Various plant species have developed morphological and/or physiological mechanisms to improve acquisition and use efficiency of mineral nutrients when grown on poor, infertile soils [79].

By such mechanisms plants achieve nutrient deficiency stress tolerance. However, sometimes the ability to adapt to stress is limited, depending on the intensity and duration of stress. For example, some nutrient stress factors, when operating for a long period, can result in inhibited growth and root physiological processes responsible for nutrient uptake with the consequence of an impaired capability for adaptation and stress tolerance. A wide range of morphological, anatomical and physiological plant traits can be responsible for variations in response to nutrient stress within a plant species [50].

The differences in their ability to grow in soils of low phosphorus status might be attributed to several factors including differential influx rates, root length/shoot weight ratios or root system morphology and root hair density. Among the more important factors is the role of root exudates in making available soil P fixed in forms such as iron or aluminium phosphate [50]. Table 3 illustrates the dominant global soil stresses.

Table 3. Dominant soil stresses [3]

Dominant soil stress	**Global land area**	
	Million km2	% of total
Continuous moisture stress	36.5	27.9
Continuous low temperatures	21.8	16.7
Seasonal moisture stress	10.3	7.9
Low nutrient-holding capacity	7.8	6.0
High aluminium	4.1	3.1
Low moisture and nutrient status	3.5	2.7
Low water holding capacity	3.4	2.6
Other stresses	27.2	20.8
Total	130.6	100.0

Chapter 5

THE IMPORTANCE OF NITROGEN MINERAL FERTILIZATION

ROLE OF NITROGEN IN PLANTS

Plants are surrounded by the nitrogen (N) in our atmosphere. Every acre of the earth's surface is covered by thousands of pounds of this essential nutrient, but because atmospheric gaseous nitrogen is present as almost inert nitrogen (N_2) molecules, this nitrogen is not directly available to the plants that need it to grow, develop and reproduce. Despite nitrogen being one of the most abundant elements on earth, nitrogen deficiency is probably the most common nutritional problem affecting plants worldwide [22].

Healthy plants often contain 3-4% nitrogen in their above-ground tissues. These are much higher concentrations than those of any other nutrient except carbon, hydrogen and oxygen, nutrients not of soil fertility management concern in most situations. Nitrogen is an important component of many important structural, genetic and metabolic compounds in plant cells. It is a major component of chlorophyll, the compound by which plants use sunlight energy to produce sugars from water and carbon dioxide (i.e., photosynthesis). It is also a major component of amino acids, the building blocks of proteins. Some proteins act as structural units in plant cells while others act as enzymes, making possible many of the biochemical reactions on which life is based. Nitrogen is a component of energy transfer compounds, such as ATP (adenosine triphosphate), which allow cells to conserve and use the energy released in metabolism. Finally, nitrogen is a significant component of nucleic acids such as DNA (deoxyribonucleic acid), the genetic material that allows cells (and eventually whole plants) to grow and reproduce [22].

In general, N (nitrogen) is scarce under natural conditions and organisms quickly use up all that is available at any given time. In other words, under natural conditions, N is a limiting growth factor in nature. [42] Soil nitrogen exists in three general forms—organic nitrogen compounds, ammonium (NH_4^+) ions, and nitrate (NO_3^-) ions. At any given time, 95-99% of the potentially available nitrogen in the soil is in organic forms, either in plant and animal residues, in the relatively stable soil organic matter or in living soil organisms, mainly microbes such as bacteria. This nitrogen is not directly available to plants, but some can be converted to available forms by microorganisms. A very small amount of organic nitrogen may exist in soluble organic compounds, such as urea, that may be slightly available to plants [22].

The majority of plant-available nitrogen is in the inorganic (sometimes called mineral nitrogen) NH_4^+ and NO_3^- forms. Ammonium ions bind to the soil's negatively-charged cation exchange complex (CEC) and behave much like other cations in the soil. Nitrate ions do not bind to the soil solids because they carry negative charges, but exist dissolved in the soil water, or precipitated as soluble salts under dry conditions. The nitrogen in soil minerals is released as the mineral decomposes. This process is generally quite slow and contributes only slightly to nitrogen nutrition for most soils [22].

Some microorganisms can utilize atmospheric N_2 to manufacture nitrogenous compounds for use in their own cells. This process, called biological nitrogen fixation, requires a great deal of energy; therefore, free-living organisms that perform the reaction, such as azotobacter, generally fix little nitrogen each year (usually less than 20 lb nitrogen/acre), because food energy is usually scarce. Most of this fixed nitrogen is released for use by other organisms upon death of the microorganism. Bacteria such as rhizobia—infect (nodulate) the roots of, and receive much food energy from, legume—plants can fix much more nitrogen per year (some well over 100 lb nitrogen/acre) [22].

The nitrogen cycle contains several routes by which plant-available nitrogen can be lost from the soil. Nitrate-nitrogen is usually more subject to loss than is ammonium-nitrogen. Significant loss mechanisms include leaching, denitrification, volatilization, and crop removal. The nitrate form of nitrogen is so soluble that it leaches easily when excess water percolates through the soil. This can be a major loss mechanism in coarse-textured soils where water percolates freely, but is less of a problem in finer-textured, more impermeable soils, where percolation is very slow [22]. Crop removal represents a loss because nitrogen in the harvested portions of the crop plant are removed from the field completely [22].

Most plants take nitrogen from the soil continuously throughout their lives and nitrogen demand usually increases as plant size increases. A plant supplied with adequate nitrogen grows rapidly and produces large amounts of succulent, green foliage [22]. The major mechanisms for nitrogen fertilizer loss are denitrification, leaching, and volatilization. Denitrification and leaching occur under very wet soil conditions, while volatilization is most common when soils are only moist and are drying [22].

Chapter 6

ENVIRONMENTAL PROBLEMS OF NITROGEN FERTILIZERS

Commercial chemosynthetic fertilizers, especially nitrogen, are easily washed below the level of the plant's root system through the leaching of rain or irrigation. An application that is too heavy or too close to the roots of the plants may cause "burning" (actually a process of desiccation by the chemical salts in the fertilizer). As well, overly heavy applications of commercial fertilizers can build up toxic concentrations of salts in the soil, thus creating chemical imbalances. If organic materials are readily available and cheap, the expense of the commercial fertilizer should also be considered [116].

Nitrogen fertilizer use is currently linked to some water quality problems involving surface and underground drinking water supplies. Current public health regulations in the United States and other countries require water suppliers to notify customers when NO_3-nitrogen concentrations in their water exceed 10 micrograms per liter. This is because ingestion of large amounts of NO_3^- by very young infants can lead to development of a potentially fatal, but treatable, condition called methemoglobinemia. Nitrate ingestion is usually not harmful to older children or adults [22]. Nitrate is not absorbed to any appreciable extent by soil particles. Nitrogen is, therefore, the most likely fertilizer element to be leached out into surface water or groundwater, or to be lost to the atmosphere by denitrification, partly as nitrogen gas and partly as nitrous oxide, which is a greenhouse gas.

The use of chemosynthetic mineral fertiliser based fertilisation regimes has increased the susceptibility of plants to diseases (e.g., powdery mildew, lodging and *Fusarium* spp.), while the same disease remained at very low levels when organic fertility management practices were used. This supports the hypothesis that the use of highly water soluble, and therefore readily plant available, chemosynthetic mineral fertilisers (especially N-fertilisers) will increase the need for the use of pesticides and other crop protection products. This provides further evidence for the hypothesis that increased chemosynthetic mineral fertiliser use results in an increased need to use chemosynthetic pesticides and thereby indirectly affects human health [122].

Natural ecological/biological processes that increase the availability of mineral nutrients in soil (e.g. ,nitrogen fixation, and P and micronutrient uptake via symbiotic mycorrhizal fungi) and protect crops against pests and diseases (e.g., natural enemies of pests and antagonists of soil-borne diseases) were also shown to be inhibited by specific agrochemical inputs [122].

The chemosynthetic fertilizers used in conventional agriculture contain just a few minerals, which dissolve quickly in damp soil and give the plants large doses of the minerals—just at one time and often more than is needed [124].

Excessive concentrations of nitrogen and phosphorus in water bodies can result in eutrophication of slow-flowing rivers, lakes, reservoirs and coastal areas. This manifests itself through a proliferation of blue-green algae, reduced light penetration, depletion of oxygen in surface water, disappearance of benthic invertebrates, and production of toxins that are poisonous to fish, cattle and humans [125].

Soils are also at risk of eutrophication in cases where excessive nutrients deplete oxygen in the soil. The result is that the natural microorganisms cannot functioning properly and soil fertility is affected. Eutrophied soils are also sources of nitrogen gases [125].

Crop productivity in intensive agriculture has increased substantially by high inputs of soluble fertilizers and pesticides—mainly nitrogen. Under natural conditions, nitrogen is biologically bound and often limits production. Mineral nitrogen in soils may contribute to the emission of nitrous oxides and is one of the main drivers of agricultural emissions. The efficiency of fertilizer use decreases with increasing fertilization, when a great part of it is not taken up by the plant but emitted into the water bodies and the atmosphere. In summary, the emission of GHG in CO_2 equivalents from the production and the application of nitrogen fertilizers from fossil fuel amounted to approximately 480 million tonnes (one percent of total global GHG emissions) in 2007. In 1960, 47 years earlier, it was less than 100 million tonnes [126].

Within the EU, mineral (commercial) chemosynthetic fertilizers are applied to agricultural soils mainly as straight nitrogen fertilizers in the form of ammonium nitrate. Nitrogen in commercial fertilizer is particularly soluble to facilitate uptake by crops, but this also makes it vulnerable to run-off after heavy rainfall, and to leaching to groundwater. [125].

Nitrogen from any kind of fertilizer affects the amounts of vitamin C and nitrates as well as the quantity and quality of protein produced by plants. When a plant is presented with a lot of nitrogen, it increases protein production and reduces carbohydrate production. Because vitamin C is made from carbohydrates, the synthesis of vitamin C is reduced also. Moreover, the increased protein that is produced in response to high nitrogen levels contains lower amounts of certain essential amino acids such as lysine and consequently has a lower quality in terms of human and animal nutrition. If there is more nitrogen than the plant can handle through increased protein production, the excess is accumulated as nitrates and stored predominately in the green leafy part of the plant. Because organically managed soils generally present plants with lower amounts of nitrogen than chemically fertilized soils, it would be expected that organic crops would have more vitamin C, less nitrates and less protein but have a higher quality than comparable conventional crops [124].

Potassium fertilizer can reduce the magnesium content and, indirectly, the phosphorus content of at least some plants. When potassium is added to soil, the amount of magnesium absorbed by plants decreases. Because phosphorus absorption depends on magnesium, less phosphorus is absorbed as well. Potassium is presented to plants differently by organic and conventional systems. Conventional potassium fertilizers dissolve readily in soil water ,presenting plants with large quantities of potassium, while organically-managed soils hold moderate quantities of both potassium and magnesium in the root zone of the plant. Given the plant responses just described, it would be expected that the organic crops would contain larger amounts of magnesium and phosphorus than comparable conventional crops [124].

Chapter 7

THE IMPORTANCE OF PHOSPHOROUS MINERAL FERTILIZATION

Fertilizers are an essential component of industrial agriculture. Plants extract soil nutrients during their growth; harvest, transport, and consumption of crops transfers soil nutrients into animal-including human-bellies. Most agricultural practices remove nutrients faster than they are replenished by natural processes of soil formation; consequently potassium, nitrogen, phosphorus and micronutrient fertilizers are purchased and applied by most farmers [36].

Global food production depends upon availability of phosphate rock, a nonrenewable fertilizer ore that is in diminishing global supply [36]. Phosphorus is often a limiting nutrient in natural ecosystems. That is, the supply of available phosphorus limits the size of the population possible in those ecosystems. When the soil is deficient in a particular nutrient element and the crop grown on such soil shows visible deficiency symptoms, there is an urgent need to correct it to improve the crop production [41].

P is second only to N as the most limiting element for plant growth [108].

Phosphorous (P) fulfills a series of functions in plant metabolism and is one of the essential nutrients required for plant growth and development. It has functions of a structural nature in macromolecules such as nucleic acids and of energy transfer in metabolic pathways of biosynthesis and degradation [62]. Next to nitrogen, phosphorous is the most abundant nutrient contained in microbial tissue, making up as much as 2% of the dry weight [55].

Phosphorous occupies a key place among nutrients because of its relative scarcity and its essential role in all life forms. P is a major constraint on food and fibre production in many parts of the world. Therefore, an economical supply of P is a necessity for secure production in agriculture and forestry. An inefficient way to satisfy this P demand is by the exclusive use of inorganic fertilizers, which rock phosphate minerals normally contain heavy metal contaminations and which often have relatively low-use efficiencies in the field [93]. The elemental state of phosphorus is highly bondable, meaning it readily combines with several other elements, which is why it is predominantly found in phosphate form. As such, phosphorus molecules bond with soil molecules and are not available for plant use until the soil is saturated or until there are no longer any soil particles for the phosphorus to bond with. The result has been a great demand for artificial fertilizers to boost nutrient content in less fertile soil [23].

The ability to obtain the minimum phosphate concentration needed for reasonable growth depends in part on the microbial environment of the root. Under some conditions, the rhizosphere microorganisms will contain species that will solubilise phosphate from compounds of very low solubility and so increase the phosphate supply to the plant. Due to the slow rate of diffusion of phosphate, the phosphate supply to a plant depends much more on the size of the root system, the density of its root hairs and the intensity of its ramifications through the soil than does the supply of the most nutrients [55].

The amount of P in plants ranges from 0.05% to 0.30% of total dry weight. The concentration gradient from the soil solution to the plant cell exceeds 2,000-fold, with an average free P of one m in the soil solution [9].

However, although bound P is quite abundant in many soils, it is largely unavailable for uptake [9, 83].

Crop yield on 40% of the world's arable land is limited by P availability. P is unavailable because it rapidly forms insoluble complexes with cations and is incorporated into organic matter by microbes. The acid-weathered soils of the tropics and subtropics are particularly prone to P deficiency and aluminum (Al) toxicity [106]. In intensive agriculture, a grain crop yield of seven metric tons Ha^{-1} requires the addition of 90 to 120 kg P Ha^{-1}. However, even under adequate P fertilization, only 20% or less of that applied is removed by the first year's growth [107].

A low P concentration in the soil solution is usually adequate for normal plant growth. For example, Barber suggested that 0.2 ppm P was adequate for optimum growth. However, for plants to absorb the total amounts of P required to produce good yields, the P concentration of the soil solution in contact with the roots requires continuous renewal during the growth cycle.[7].

Chapter 8

THE PROBLEM WITH THE EFFECTIVENESS OF MINERAL FERTILIZERS

The efficiency of chemosynthetic fertilizers might be lower than 50% either due to mismanagement or due to reactions of the fertilizers with soil constituents. only a fraction of the fertilizer applied to the soil is taken up by the crop; the rest either remains in the soil or is lost through leaching, physical wash off, fixation by the soil, or release to the atmosphere through chemical and microbiological processes. This low uptake and utilization of nutrients out of the added fertilizers results in lower production and appearance of nutrient stress symptoms. The poor efficiency of added fertilizers is mainly due to imbalance fertilization or improper management practices [41].

There are three main types of chemosynthetic nitrogen fertilizers in which the nitrogen in the fertilizer is found in the chemical form of ammonium, urea and nitrate. Nitrogen in nitrate-form fertilizers are soluble and highly mobile in soil. They are prone to various kinds of losses: leaching beyond the root zone in the light textured soils, denitrification in waterlogged conditions or short time of over irrigation. Ammonium type fertilizers usually transform in the soil within a week or two to nitrate form and the same applies to nitrate fertilizers [41].

Chemosynthetic phosphate fertilizers, which are highly reactive, are fixed in soil and become immobile. The assessment of the profitability of P fertilisation is generally based on results of agronomic trials. Fertilizer trials are predominantly carried out under optimum or near-to-optimum management conditions (soil preparation, sowing time, weeding) which do not reflect on-farm conditions, in particular when farmers rely on manual labour. For instance, sowing dates are normally distributed over several weeks so that conditions for germination and emergence in many cases are sub-optimal. Therefore, plant establishment and responses to fertilizer are lower on average than those of trial results [93]. Table 4 illustrates the different type of phosphate fertilizers.

P is absorbed mainly during the vegetative growth and, thereafter, most of the absorbed P is re-translocated into fruits and seeds during reproductive stages. P-deficient plants exhibit retarded growth (reduced cell and leaf expansion, respiration and photosynthesis), and often a dark green colour (higher chlorophyll concentration) and reddish coloration (enhanced anthocyanin formation) [62].

Table 4. Different type of phosphate fertilizers

Source	% P_2O_5	Water soluble %
Normal superphosphate	16	90
Concentrated superphosphate	44-52	95-98
Monoammonium phosphate	48-61	100
Diammonium phosphate	48-53	100
Ammonium polyphosphate	34-37	100
Phosphoric acid	55	100
Super phosphoric acid	76-85	100

The concentrations of inorganic P in soil solutions are low because P is rapidly sorbed to soil colloids [69]. Where a water-soluble phosphorus (WSP) fertilizer is applied to the soil, it reacts rapidly with the soil compounds. The resulting products are sparingly soluble P compounds and P adsorbed on soil colloidal particles [24] and only up to 45% of phosphorus fertilizers taken up by crops [89].

Factors affecting phosphorous availability are as follows [42]:

1) Amount of clay — increased clay content results in greater retention of P by the soil.
2) Type of clay — kaolinite and iron oxides retain more P than 2:1 clays.
3) Soil pH — determines the form of P, and the levels of Al, Fe and Ca in soil solution. Also affects pH dependent charge.
4) Soil P levels — soils very high in P will eventually saturate binding sites and tend to release higher concentrations of P to soil solution.
5) Temperature — at low and high temperatures, ability of the plant to remove P is decreased
6) Compaction — limits root penetration and volume of the soil contacted by the root system
7) Aeration — poorly aerated and poorly drained soils can limit root system development.
8) Moisture — low moisture limits movement of P and affects crop development; excess moisture affects rooting and limits P uptake.
9) Time and method of application — Increasing the time of contact increases the amount of P retained by the soil in unavailable forms. On soils with high P retention, banded applications may provide more available P to the crop when soil P levels are low. As soil fertility levels increase, the banding advantage decreases, although starter fertilizers frequently result in improved yields due to improved early season development in some crops. Broadcast applications are more rapid and cheaper to apply, can provide larger rates without crop injury, and can result in better mixing within the entire root zone which over time can increase rooting depth.
10) Solubility and/or particle size — Retention tends to increase as the solubility increases and as particle size of the applied P decreases.

11) Incorporation vs. surface application — Surface applied P is usually the least efficient method of placing P. Incorporation throughout the plow layer leads to improve rooting depth and more effective uptake over time. In no-till situations, incorporation is not possible. Building good P levels prior to initiating a no-till system is strongly recommended. As residues build on the soil surface, natural P cycling in the soil by animals, and movement through macropores, surface applications of can supply the needs of the crop.
12) Other nutrients — although P availability is little affected by the presence of other essential nutrients within normal ranges, the presence these nutrients can stimulate P uptake.

Management and research must aim at nutrient and organic matter cycling in combination with sufficient fertilizer use to avoid "nutrient mining" in agriculture and forestry.

Many natural ecosystems and low-input farming systems have adapted to low P supply by recycling P from litter and soil organic materials. Increases in productivity or exports of commodities require external nutrient inputs if they are not to cause a decline in fertility [95].

Phosphorus is a major constraint on food and fibre production in many parts of the world. Therefore, an economical supply of P is a necessity for a secure production in agriculture and forestry. An inefficient way to satisfy this P demand is by the exclusive use of inorganic fertilisers, which normally are expensive imports, and which often have relatively low use efficiencies in the field [95].

Importance of Nutrient Balance

A positive net mass balance indicates the amount of residual nutrient that may remain in the soil or be lost to the air, carried by water runoff into surface-water systems, or carried by percolating water into ground water. A negative net balance indicates that the amount of nutrient removed from the field through the harvested crop exceeds the amount of nutrient applied, with the difference coming from nutrients stored in the soil or available through precipitation. Continued negative balances mine or deplete nutrients in soil, disrupt the soil ecosystem, and can damage soil productivity [39].

In areas with high soil erosion and run off, the high residual balance of phosphorus could contribute to water quality problems and require improved management. Phosphorus is more stable than nitrogen and more likely to remain in the soil with less loss to the environment unless the soil itself erodes away. Because of this greater stability, and to reduce costs, many farmers apply extra phosphorus one year then skip a year or more [39].

Chapter 9

PHOSPHATE ROCK IS A FINITE, NONRENEWABLE NATURAL RESOURCE

The current world production of food is dependent on large amounts of fossil fuels (fertilizer production) and finite reserves of phosphate rock. Canada has no significant mineable sources of phosphate for inorganic fertilizer production and is entirely dependent on imported sources, mainly the United States.

PR is a finite, non-renewable natural resource. Geological deposits of different origin are found throughout the world. Currently, few PR deposits are mined, and about 90 percent of world PR production is utilized by the fertilizer industry to manufacture P fertilizers, with the remainder being used to manufacture animal feeds, detergents and chemicals [25].

In fact, if viewed from a global perspective, phosphate rock is not a purchased input, but represents a nonrenewable with finite bounds of particular importance in feeding a burgeoning world population [10] and for which there is no substitute [36].

Most of the world's farms do not have or do not receive adequate amounts of phosphate. Feeding the world's increasing population will accelerate the rate of depletion of phosphate reserves. Rocks containing phosphate have been discovered and are being mined at minimal cost. But resources are limited, and phosphate is being dissipated. Future generations ultimately will face problems in obtaining enough to exist [1]. P is a vitally important resource and its future use in agriculture will be impacted by declining availability and increased cost.

While most economic studies indicate that farmer response to fertilizer price changes is fairly inelastic (i.e., relatively unresponsive to price changes), farmers can make some adjustments in fertilizer use when faced with dramatically higher prices [114].

Table 5. Average U.S. farm prices of selected phosphate fertilizers (dollars per ton) [103]

Year	Month	Super-phosphate 20% phosphate	Super-phosphate 44-46% phosphate	Diammonium phosphate (18-46-0)
1960	September	38.0	80.1	NA
1970	September	46.9	76.2	95.1

Table 5. (Continued)

Year	Month	Super-phosphate 20% phosphate	Super-phosphate 44-46% phosphate	Diammonium phosphate (18-46-0)
1974	September	104.0	188.0	228.0
1980	May	132	251	298
1990	April	NA	201	219
1995	April	NA	234	263
2000	April	NA	233	240
2001	April	NA	236	244
2002	April	NA	221	227
2003	April	NA	243	250
2004	April	NA	266	276
2005	April	NA	299	303
2006	April	NA	324	337
2007	April	NA	418	442

Table 6. Fertilizer price indexes [103]

	Index of prices	
Year	Paid by farmers for fertilizer	Received by farmers for all crops
1976	74	86
1980	96	105
1984	103	109
1990	97	101
1992	100	100
1995	105	95
2000	110	96
2001	123	99
2002	108	105
2003	124	111
2004	140	117
2005	162	112
2006	174	120

Table 7. Consumption of selected phosphate fertilizers (Metric tons) [103]

Year	Superphosphates Grades 22% and under	Superphosphates Grades over 22%	Superphosphates Diammonium phosphate (18-46-0)	Other nitrogen-phosphate grades 2
1960	510,539	399,319	20,388	NA
1968	401,347	1,109,332	1,246,953	NA
1970	312,032	1,204,566	1,514,911	NA
1976	140,343	1,211,060	3,232,501	NA
1981	112,524	1,051,653	3,721,845	2,510,250
1985	68,574	751,412	3,433,653	2,730,920
1990	13,717	579,841	3,580,222	3,065,005
2000	11,437	333,298	3,220,179	4,114,562
2004	21,421	213,702	3,580,108	4,469,690
2005	15,635	191,061	3,392,521	4,869,201
2006	8,993	173,939	2,999,062	4,925,666

Estimates on the remaining amount of phosphorus vary as do projections about how long it will take to deplete the irreplaceable resource entirely. Figures range from 60-130 years [91] and 60-90 years [94] at current market prices with diverse assumptions about the rate of production and demand, but all sources agree that continued phosphorus production will decline in quality and increase in cost. The relatively inexpensive phosphorus we use today will likely cease to exist within 50 years [23]. By some estimates, inexpensive rock phosphate reserves could be depleted in as little as 60 to 80 years. P fertilizer use increased 4- to 5-fold between 1960 and 2000 and is projected to increase further by 20 Tg^{-1} per year by 2030. [107] Between 1950 and 2000, about 1 billion metric tonnes of P has been mined. During this period, about 800 million metric tonnes of fertilizer P were applied to the earth's croplands. This has increased the standing stock of P in the upper 10 centimetres of soil in the world's croplands to roughly 1,300 million metric tonnes, an increase of 30% [23].

PR sedimentary rock, which accounts for some 85-90% of world production, contains Cd in concentrations ranging from less than 20 to more than 200 mg per kg P_2O_5 [72].

The European fertilizer industry is almost totally dependent on imports of rock phosphate from Africa, the Near East, Russia and the U.S.

There is a limit to the phosphate rock reserves that can profitably be recovered at current market prices. High-grade ores with high P_2O_5 concentration and ores of good quality, i.e., containing few contaminants, are being progressively depleted. Consequently, production costs will increase. In the context of declining quality, the levels of certain impurities may pose problems in the processing or in application. The content of heavy metal contaminants are generally higher in sedimentary than igneous rocks. To date, levels have not proved a problem except when producing higher purity phosphoric acid grades. However, certain

European countries have applied strict limits to cadmium levels in fertilisers in order to avoid contamination of farmland and crops and cadmium removal could involve further processing costs adding 2-10% to phosphate fertiliser prices [91].

Phosphate content in currently mined rocks can range from over 40% to below 5%. The mined rock is further processed to remove the bulk of the contained impurities and thus upgrade the rock. Consequently, the rock concentrate contains an increased apatite content of an improved quality. Phosphate rock can be beneficiated by many methods, and usually a combination of methods is used. In general, with lower concentration of phosphate and lower quality deposit, more waste is generated. Furthermore, more energy and chemicals are required per tonne of useful phosphate produced. Consequently, the cost for recovery and beneficiation of phosphate rock increases significantly in relation to lower grade and lower quality deposit [91].

It imperative that we begin recycling phosphorus and returning it to the soil to decrease the need for mined phosphorus as artificial fertilizer. Within a half century, the severity of this crisis will result in increasing food prices, food shortages and geopolitical rifts [23].

The critical question, when phosphate ore reserves will be exhausted, is hard to answer, however, the economic mining of phosphate ore is estimated to continue for a few decades only [93].

There is a growing concern that the world is coming close to its 'limits of growth' as the earth's natural resources are limited. P is one of the essential inputs for agricultural production and growth. Environmental quality comprises two important issues: the maintenance of soil fertility and the avoidance of environmental pollution.

Therefore, is an urgent need to reduce the loss of nutrients from agriculture, in order to slow the consumption of finite phosphate rock deposits. A potential phosphate crisis looms for agriculture in the twenty first century. Sustainable management of P in agriculture requires that plant biologists discover mechanisms in plants that enhance P acquisition and exploit these adaptations to make plants more efficient at acquiring P, develop P-efficient germplasm, and advance crop management schemes that increase soil P availability [107].

There are no high P content (15-18%) materials known than apatite minerals, which are either in phosphate rock or bone charcoal 3R-AGROCARBON formation. The P content of other materials may occur as low as 0.1-0.5 % if at all.

Organic P from manure or sludge should be comparable to P from inorganic fertilizer. Therefore, if a producer has a P recommendation for 33 kg/ha of P_2O_5, applying approximately 73kg of 18-46-0 diammonium phosphate fertilizer or six tons of 11-6-9 manure (80% available P coefficient) should provide equivalent results [78].

A significant percentage of phosphorus can be recovered by using sustainable agriculture and sanitation. This should be a priority for the global policy agenda.

Chapter 10

THE INORGANIC CONTAMINATION OF PHOSPHATE ROCK (PR) FERTILIZERS

The process to extract phosphorus is particularly damaging, as strip-mining is the most common method of extraction. The by-products of the processing of phosphorus include radioactive gypsum, commonly stored in "stacks" in the United States, and air pollution that includes fluorides. Furthermore, phosphorus deposits are often contaminated by arsenic or cadmium, which can be concentrated during processing and there is a subsequent build-up of these toxic substances in agricultural soils [23].

Once mined, rock phosphorite is crushed and acidified at an enormous scale. Eight five percent of global Sulphur is used to make sulphuric acid; 70% of this industrial acid is used to process phosphate rock into usable fertilizer. Several different acidification processes are employed to produce phosphorus-rich liquids and precipitates that are sold and applied to soils as fertilizer and animal feed supplements [36].

PR contains various metals and radionuclides as minor constituents in the ores. Varying amounts of these elements are transferred to phosphate fertilizers in production processes, and later are applied to soils with these fertilizers. Cadmium (Cd) is the heavy metal of most interest because it is potentially harmful to human health, and much attention is being given to its avenues of entry into the human food chain [93].

Cd from phosphate fertilisers poses a potentially serious threat to soil quality and, through the food chain, to human health. The exact size of the problem is hard to determine due to a lack of sufficient and reliable data. However, it is obvious that in the long term the continuing input of Cd to agricultural soils through fertilisers could lead to accumulation beyond acceptable levels. While there are other important sources of soil contamination with Cd as well (e.g., manure, compost, sewage sludge and atmospheric deposition), phosphate fertilisers are generally the major contamination sources and the most important ones as far as farmland is concerned. The European Commission is considering the possibility of using an European Union (EU)-wide charge on Cd in fertilisers so as to improve the competitive position of the 'low-Cd' product *vis-á-vis* the 'high-Cd' one. Such a charge would encourage farmers to use the low-Cd alternatives and stimulate suppliers to invest in technologies for producing them [72].

Cd occurs naturally as a contaminant in all phosphate rock, but the concentrations vary considerably depending on the origin of the material. Igneous rock or, apatite (found in the former Soviet Union, Finland, South Africa and South America), has low concentrations of Cd (often less than one mg per kg P_2O_5) [72].

Decadmiation is currently too expensive and low-cost technologies are not yet fully developed. In some cases technical restrictions impede the use of either magmatic rock or decadmiation [72].

Obviously, the flow of cadmium to agricultural soils can also be reduced by applying lower amounts of phosphate fertiliser and/or substituting fertilisers by other products containing phosphorus, e.g., manure. While this may be a feasible option in certain cases, it does not take away the need to minimise the amount of cadmium in fertilisers. Moreover, products such as manure, compost and sewage sludge may also contain substantial amounts of cadmium and/or other contaminants [72].

There are still no agreed safe Cd limits for soils, but it appears that concern about Cd build-up in soils may be warranted only where several critical factors combine, i.e., on acid soils with low cation exchange and low P fertility, to which significant P fertilisation is applied as low-grade fertilizer or rock phosphate, particularly to Cd accumulators like leafy vegetable crops. In the natural environment, P is supplied through the weathering and dissolution of rocks and minerals with very low solubility. Therefore, P is usually the critical limiting element for plant and animal production. Present legislative limits in the EU member states vary from 21.5 mg Cd/kg P_2O_5 while the existing Cd concentrations in fertilizers is up to 58 mg Cd/kg P_2O_5. In order to decrease the Cd accumulation in agricultural soils, a 40 mg upper limit Cd/kg P_2O_5 concentration is proposed in P fertilizers by 2010, and 20 mg Cd/kg P_2O_5 by 2015. In order to maintaining the status quo, the industry proposes to accept an upper limit of Cd in phosphate fertilizer of 60 mg Cd/kg P_2O_5. A limit variation for Cd in phosphate fertilizers might also be derived based on a risk assessment approach. However, rock phosphate resources with low Cd content are much less available than expected in the past, meaning that meeting an upper limit for the annual average for each manufacturer might be problematic to achieve [51].

The Cd content of rock phosphate resources varies considerably from source to source, revealing as much as a hundred-fold difference when extremes are compared. While the purchase of rock with a low Cd content is clearly desirable, it should be noted that there are insufficient quantities available in the world to meet food production needs. While the 1988 EU directive on Cd emissions to water recognised the desirability of reducing inputs of Cd to soils as part of an overall strategy for reducing environmental contamination by Cd, it also acknowledged that by 2007 there was still no economically viable process to remove Cd from phosphate rock.

A side effect of sustained heavy P fertilizer additions can be the accumulation and introduction into the human food chain of heavy metal contaminants contained in fertilizers. Phosphate fertilizer use has caused small but significant increases soil Cd levels. These inputs are of a magnitude similar to those from the atmosphere in industrialised countries. There are still no agreed safe Cd limits for soils, but it appears that concern about Cd build-up in soils may be warranted only where several critical factors combine, i.e., on acid soils with low cation exchange and low P fertility, to which significant P fertilisation is applied as low grade fertilizer or rock phosphate, particularly to Cd accumulators like leafy vegetable crops, and where the produce is the main source of local food consumption.

Availability of fertilizer Cd to plants is related to its solubility in the P fertilizer as well as to soil factors such as pH, cation exchange capacity, and clay and organic matter contents.

Plants vary in both their capacities to absorb and to translocate Cd from vegetative tissues to grain [93].

Table 8 presents the average heavy metal concentrations in phosphate rock deposits and estimated input to soil by P fertilizers [53].

Table 8. Average heavy metal concentrations in phosphate rock deposits and estimated input to soil by P fertilizers

	Heavy metal concentration								
PR Deposit	As	Cd	Cr	Cu	Pb	Hg	Ni	V	Zn
	---------------------- mg kg^{-1} of PR --------------------------								
Russia (Kola)	1	0.1	13	30	3	0.01	2	100	19
USA	12	11	109	23	12	0.05	37	82	204
South Africa	6	0.2	1	130	35	0.06	35	3	6
Morocco	11	30	225	22	7	0.04	26	87	261
Other N. Africa	15	60	105	45	6	0.05	33	300	420
Middle East	6	9	129	43	4	0.05	29	122	315
Avg. of 91%	---------------------- mg kg^{-1} of PR --------------------------								
of PR reserves	11	25	188	32	10	0.05	29	88	239
	-------------------------- mg kg^{-1} of P --------------------------								
	71	165	1226	209	66	0.29	189	578	1561
Applied with	--------------------------g ha^{-1} ----------------------------								
20 kg P ha^{-1}	1	3.3	25	4	1	0.01	4	12	31
Tolerable limit	------------------------ mg kg^{-1} of soil ------------------------								
	-	2	100	100	100	2	50	50	300

Cd is a toxic heavy metal. Humans exposed to high levels over a number of years may develop chronic Cd poisoning. The most recent concern is the possibility that a growing number of people are being exposed to Cd through its uptake in their diet. This has now brought phosphate fertilizers to the centre of the debate. The major concern is that Cd slowly accumulates in the kidney causing irreversible damage, where the concerns are as follows:

- Cd does not have an environmental half-life but persists indefinitely in the environment.
- Cd is removed from the human body only slowly (half-life on the order of 20 years).

- Cd causes irreversible damage to the target organ, the kidney, and significantly increases the human health risk.

Maximum levels of Cd in foodstuffs are already set by the Commission's regulation EC 466/2001 of 8 March of 2001. Sedimentary rock, which accounts for some 85-90% of world production, contains Cd in concentrations ranging from less than 20 to more than 200 mg per kg P^2O^5. De-cadmiation is currently too expensive and low-cost technologies are not yet fully developed. In some cases technical restrictions impede the use of either magmatic rock or de-cadmiation [72].

Concentrations of other heavy metal and radionuclide contaminants in P fertilizers vary considerably, depending on the PR source. Some PR source metals of possible significance are arsenic (As), chromium (Cr), lead (Pb), mercury (Hg), nickel (Ni), and vanadium (V). The main radionuclide contaminants in PR are uranium (U), radium (Ra), and thorium (Th) [93].

Typical levels of gamma radiation from the PR deposits range from 5 to 30 kBq kg^{-1} P [53].

The median contents of PR in the U.S. were reported as 59 mg kg^{-1} of U, 8 mg kg^{-1} of Th, and 18 mg kg^{-1} of Ra [66]. The wide use of phosphate fertilizer will certainly cause high radiological environmental pollution. Evidence of very high alpha activity in some soil samples may indicate excessive use of phosphate fertilizer in agricultural lands. As phosphate fertilizers are in the form of fine powder, precautionary measures need to be taken to minimize the level of airborne dust during processing, transportation, and application [30].

Chapter 11

THE ENVIRONMENTAL RISK OF THE PHOSPHATE FERTILIZER INDUSTRY

Nearly all chemosynthetic phosphate fertilizer is produced by treating phosphate rock with sulfuric acid to produce phosphoric acid, which is further processed into various phosphatic fertilizer materials such as superphosphates and ammonium phosphates. The United States has become the world.s largest phosphate fertilizer exporter [39].

Approximately 1.7 tons of phosphate rock and about 1.3 tons of sulfuric acid are required to produce a ton of diammonium phosphate [39].

Fluoride has been one of the largest environmental liabilities of the phosphate industry. The source of the problem lies in the fact that raw phosphate ore contains high concentrations of fluoride, usually between 20,000 to 40,000 parts per million (equivalent to two to four percent of the ore). When this ore is processed into water-soluble phosphate (via the addition of sulfuric acid), the fluoride content of the ore is vaporized into the air, forming highly toxic gaseous compounds (hydrogen fluoride and silicon tetrafluoride). In the past, when the industry had little, if any, pollution control, the fluoride gases were frequently emitted in large volumes into surrounding communities, causing serious environmental damage [16].

To make one pound of commercial fertilizer, the phosphate industry creates 5 pounds of contaminated phosphogypsum slurry (calcium sulfate). This slurry is piped from the processing facilities up into the acidic wastewater ponds that sit atop the mountainous waste piles known as gypsum stacks. After phosphate rock has been reacted with sulfuric acid to produce phosphoric acid, the precipitated gypsum is removed with filters and pumped in slurry form to an impoundment where it is allowed to settle. As the gypsum accumulates, a small dragline removes some of the gypsum for raising the height of the dikes. By this process the gypsum settling impoundment, or "stack" as it is often called, increases in elevation. As a stack grows in height (up to 200 feet) the area of the settling impoundment decreases until a point is reached where the pond capacity becomes too small and the pumping height requires too much energy [64].

It is estimated that more than 22 million tonnes of phosphoric acid (as P_2O_5) are produced annually worldwide, generating in excess of 110 million tonnes of gypsum by-product. The free water in the gypsum cake off the filters is highly acidic, having a pH as low as 1.0. While commercial uses, in agriculture and in manufacturing gypsum board and Portland cement, consume less than a few percent of this by-product, the vast majority is disposed of on land in gypsum stacks or is discharged into water bodies [5].

Since the mid-1980's, the annual production rate of phosphogypsum has been in the range of 40 to 47 million metric tons per year. The total amount generated in the United States from 1910 to 1981 was about 7.7 billion metric tons [105]. According to the EPA, 32 million tons of new gypsum waste is created each year by the phosphate industry in central Florida alone. (Central Florida is the heart of the US phosphate industry). The EPA estimates that the current stockpile of waste in Central Florida's gypsum stacks has reached nearly one billion metric tons. (The average gypsum stack takes up about 135 acres of surface area — equal to about 100 football fields — and can go as high as 200 feet [16].

The pH of the phosphogypsum is very acidic: pH 2.8. The acidity is due to residual phosphoric and sulfuric acids plus hydrofluoric and fluosilicic acids [64].

Fluoride emissions from phosphogypsum pond water surfaces are primarily HF and have been measured to be in the range of 0.01 to 0.10 kg/hectare/day. Fluoride emissions from dry gypsum surfaces are associated with fugitive gypsum dust. Vegetation very close to a gypsum stack can contain elevated levels of fluorides which, if ingested by cattle for years, can cause fluorosis [5].

Phosphate rock, which is processed to make phosphoric acid, contains concentrations of naturally occurring radioactive elements (radionuclides). Even high-grade ores, which contain about 70% calcium phosphate, also contain a large number of impurities-such as calcium fluoride, chlorides, chromium, rare earths, and radionuclides. At the end of the production process, the radionuclides end up in the phosphogypsum [105].

The phosphogypsum, separated from the phosphoric acid, is in the form of a solid/water mixture (slurry) which is stored in open-air storage areas known as stacks. The stacks form as the slurry containing the by-product phosphogypsum is pumped onto a disposal site. Over time, the solids in the slurry build up and a stack forms. The stacks are generally built on unused or mined-out land on the processing site. The surface area covered by stacks ranges from about 5 to 740 acres. The height ranges from about 10 to 200 feet. In 1989, the total surface area covered by stacks was about 8,500 acres. More than half that acreage is in Florida [105].

Phosphogypsum is virtually useless, and is disposed of in large, aboveground stacks, or piles. A total of 63 phosphogypsum stacks were identified nationwide in 1989, in 12 different states. Two-thirds of these stacks were located in Florida, Texas, Illinois, and Louisiana. In central Florida, one of the major phosphoric acid producing areas, the industry generates about 32 million metric tons of phosphogypsum each year, which equals the combined weight of approximately 6.4 million elephants.[105].

It is sort of a misnomer, however, to call these stacks "gypsum" stacks. Indeed, if the stacks were simply gypsum, they probably wouldn't exist, as gypsum can be readily sold for various purposes (e.g., as a building material). What can't be readily sold, however, is radioactive gypsum, which is about the only type of gypsum the phosphate industry has to offer. The source of the gypsum's radioactivity is the presence of uranium, and uranium's various decay products (i.e., radium), in raw, phosphate ore [16].

There is a natural and unavoidable connection between phosphate mining and radioactive material. It is because phosphate and uranium were laid down at the same time and in the same place by the same geological processes millions of years ago. They go together. If you mine phosphate, you get uranium. While uranium and its decay-products naturally occur in phosphate ore, their concentrations in the gypsum waste, after the extraction of soluble phosphate, are up to 60 times greater [16].

The concentrations of uranium and radium-226 in phosphogypsum samples taken in central Florida were about 10 times the background levels in soil for uranium and 60 times the background levels in soil for radium-226.

The radium-226 concentration in phosphogypsum varies significantly at different sampling locations on a single stack and also in phosphogypsum from different stacks within the same geographical area [105].

EPA is currently "weighing whether to classify the gypsum stacks as hazardous waste under federal statutes, which would force the industry to provide strict safeguards" (to nearly one billion tons of waste) [16].

Radionuclides that are small particles (i.e., radionuclide dust) can become airborne as wind-blown dust or as dust thrown up into the air by cars and trucks. The radionuclides, uranium and radium-226, are present in the phosphogypsum and can become airborne. Once these radionuclides are in the air, people and animals can breathe them and they can settle out onto ponds and agricultural areas. Radon-222, a decay product of radium-226, is a gas and so may become airborne by diffusing into the air. The EPA has determined, however, that the risks associated with stacking phosphogypsum are in line with acceptable risk practices [105].

Phosphogypsum contains some trace metals in concentrations that the EPA believes may pose a chemical hazard to human health and the environment. Analysis of samples from various facilities contained arsenic, lead, cadmium, chromium, fluoride, zinc, antimony, and copper at concentrations that may pose significant health risks. The concentrations of these contaminants vary by more than three orders of magnitude among samples taken from various locations. These trace metals may also be leached from phosphogypsum and migrate to nearby surface and groundwater resources [105].

Only two uses (for the gypsum) are permitted: limited agricultural use and research. Other uses may be proposed, but otherwise the phosphogypsum must be returned to mines or stored in stacks [16].

While the presence of uranium-decay products makes gypsum a tough sell for the phosphate industry, the uranium has, at various times, presented the industry with a business opportunity of its own. One of the lesser-known facts about the phosphate industry is that its processing facilities have produced and sold sizeable quantities of uranium. In 1997, just two phosphate plants in Louisiana produced 950,000 pounds of commercial uranium, which amounted to roughly 16% of the domestically produced uranium in the U.S. In 1998, the same two plants produced another 950,000 pounds, but due to declining market prices for uranium, both plants have since ceased production. If market prices improve, however, four U.S phosphate plants two in Louisiana and two in Florida) would have the capacity to produce a combined 2.75 million pounds of uranium per year, according to the Department of Energy (DOE) [16].

Resting atop the phosphate industry's gypsum piles are highly-acidic wastewater ponds, littered with toxic contaminants including fluoride, arsenic, cadmium, chromium, lead, mercury, and the various decay-products of uranium. This combination of acidity and toxins makes for a poisonous, high-volume cocktail which, when leaked into the environment, wreaks havoc in waterways and fish populations [16].

SUMMARY OF PROBLEMS WITH RP

- The economically minable low heavy metal-contaminated total rock phosphate of the world is sufficient for up to another 25 years.
- VERY IMPORTANT: the world population and food consumption + energy use level is under acceleration. Therefore, these calculated numbers are not valid, as the P supply issue is much worse than the worst case ever calculated.
- The Cd, Uranium and other heavy metal issues suggest that the accessible low-CD RPs will be exhausted within less than a decade, again calculated with today's consumption level.
- If world population and food consumption + energy use level is accelerating faster (for which there are already very clear signs in India, China, Indonesia, the Philippines and Russia), then agrobusiness will also accelerate proportionally, which means that the low-Cd RP deficit is expected to be a reality as soon as the 2010s.
- If very significant investments are made in RP production and processing and by extension RP access is prolonged, but the investments will have to be in billions and billions, practically almost a new industry.
- There are ongoing technology developments to remove Cd from RP by chemical processes, but these issues will be extremely costly and industrial technically is not yet available.
- The transport energy cost throughout the world is also exploding, which is also accelerating RP prices, especially the low Cd RPs imported from long distances.

Chapter 12

Greenhouse Gas Emission from Intensive Agricultural Activities

Agricultural systems contribute to carbon emissions through several mechanisms: the direct use of fossil fuels in farm operations, the indirect use of embodied energy in inputs that are energy-intensive to manufacture (e.g., fertilizers), and the cultivation of soils resulting in the loss of soil organic matter.

Naturally occurring greenhouse gases include water vapour, carbon dioxide (CO_2), methane (CH_4), nitrous oxide (NOx), and ozone (O_3). Several classes of halogenated substances that contain fluorine, chlorine, or bromine are also greenhouse gases, but they are, for the most part, solely a product of industrial activities.

There are also several gases that do not have a direct global warming effect but indirectly affect terrestrial and/or solar radiation absorption by influencing the formation or destruction of greenhouse gases, including tropospheric and stratospheric ozone. These gases include carbon monoxide (CO), oxides of nitrogen (NOx), and non-CH_4 volatile organic compounds (NMVOCs). Although the direct greenhouse gases CO_2, CH_4, and nitrogen gas complexes occur naturally in the atmosphere, human activities have changed their atmospheric concentrations [104].

While the physics of the greenhouse effect are similar for all gases, each gas differs in its overall effect on the earth's radiation balance, depending on the concentration of the gas, its residence time in the atmosphere, and its physical properties with respect to absorbing and emitting radiant energy. A common measure, termed the global warming potential (GWP), is used to equate the effect of different greenhouse gases on a mass basis. By convention, the effect of carbon dioxide (CO_2) is assigned a value of one (1) and the warming potential of other gases are expressed relative to that standard, i.e., on a CO_2–equivalent basis [73].

The GWP (global warming potentials) of a greenhouse gas is defined as the ratio of the time-integrated radiative forcing from the instantaneous release of 1 kilogram (kg) of a trace substance relative to that of 1 kg of a reference gas (see Table 9) [43].

One tonne of nitrous oxide is deemed to have the same warming effect as 310 tonnes of CO_2, and one tonne of methane is deemed to have the warming effect of 21 tonnes of CO2.

The reference gas used is CO_2, and, therefore, GWP-weighted emissions are measured in teragrams of CO_2 equivalent (Tg CO_2 Eq.). One teragram is equal to 1,012 grams or one million metric tons [104].

Table 9. Global warming potentials of direct greenhouse gases [44]

Gas	GWP
CO_2	1
CH_4	21
NOx	310

Agricultural systems contribute to carbon emissions through the direct use of fossil fuels in food production, the indirect use of embodied energy in inputs that are energy intensive to manufacture, the cultivation of soils and/or soil erosion resulting in the loss of soil organic matter [6]. In 2006, the agricultural sector was responsible for emissions of 454.1 teragrams of CO_2 equivalent (Tg CO_2 Eq.), or six percent of total U.S. greenhouse gas emissions. Methane (CH_4) and nitrous oxide were the primary greenhouse gases emitted by agricultural activities. CH_4 emissions from enteric fermentation and manure management represent about 23 percent and seven percent of total CH_4 emissions from anthropogenic activities, respectively. Of all domestic animal types, beef and dairy cattle were by far the largest emitters of CH_4. Rice cultivation and field burning of agricultural residues were minor sources of CH_4. Agricultural soil management activities such as fertilizer application and other cropping practices were the largest source of U.S. NOx emissions, accounting for 72 percent. Manure management and field burning of agricultural residues were also small sources of nitrogen emissions [104].

Nitrous oxide (NOx) and methane (CH_4) emissions result from both crop and livestock operations and account for approximately 80 percent of U.S. agricultural greenhouse gas emissions on a GWP basis [73].

Nitrogen oxides (NOx), are emitted from agricultural soils and through combustion [19]. Tropospheric ozone, a component of smog that impacts human health, agricultural crops and natural ecosystems has increased. As much as 35% of cereal crops worldwide are exposed to damaging levels of ozone [14]. NOx from agroecosystems can be transported atmospherically over long distances and deposited in terrestrial and aquatic ecosystems. This inadvertent fertilization can cause eutrophication, loss of diversity, dominance by weedy species and increased nitrate leaching or NOx fluxes [110].

According to current projections, total greenhouse gas emissions from agriculture are expected to reach 8.3 Gt CO_2 equivalents per year in 2030, compared to the current level of approximately six Gt CO_2 equivalents annually [119]. Agricultural NOx emissions are projected to increase by 35-60% up to 2030 due to increased nitrogen fertilizer use and increased animal manure production [120].

Emissions of nitrous oxide are directly linked to the concentration of easily available mineral nitrogen in soils. High emission rates are detected directly after fertilization and are highly variable. Denitrification is additionally enhanced in compacted soils. According to IPCC, 1.6% of nitrogen fertilizer applied is emitted as nitrous oxide [121].

Nitrous oxide constitutes the largest agricultural source of GHG emissions in terms of warming potential (48 percent), and almost 70 percent of total U.S. nitrous oxide emissions are from soils. The best option for reducing these emissions is to use fertilizers more efficiently; adoption of best fertilization practices could reduce agricultural NO_x emissions by

30 to 40 percent [18]. Because NO_x emissions are strongly influenced by the availability of nitrogen in soil, some NO_x emissions are an unavoidable consequence of maintaining highly productive crop and pastureland. However, improved control of the amount, timing, and placement of fertilizer can minimize these emissions [73]. Greater than 50 percent of the major cropland area in the United States is rated as having high nitrogen balances, resulting in soils highly susceptible to losses of NO_x to the atmosphere and nitrate (NO_3^-) to water bodies [102]. Fall application of fertilizer (motivated by lower fertilizer costs in the fall and to save time during spring planting) results in high concentrations of mineral nitrogen remaining in soils over a several-month period with no plant uptake, making that nitrogen vulnerable to losses [73].

Table 10. Selected U.S. greenhouse gas (GHG) emissions from agriculture (Tg CO_2 Eq.) Source: [104]

Gas/Source	1990	2000	2006
CH_4 (total)	606.1	574.3	555.3
CH_4 (agricultural sector emission)	165.7	171.7	174.4
Percentage of total CH_4 emission (%)	27%	30%	31%
Enteric fermentation	126.9	124.6	126.2
Manure management	31.0	38.8	41.4
Rice cultivation	7.1	7.5	5.9
Field burning of agricultural residues	0.7	0.8	0.8
NOx (total)	383.4	385.9	367.9
NOx (agricultural sector emission)	281.8	276.3	279.8
Percentage of total NOx emission (%)	74%	72%	76%
Agricultural soil management	269.4	262.1	265.0
Manure management	12.1	13.7	14.3
Field burning of agricultural residues	0.4	0.5	0.5
Total CH_4 + NOx	989.5	960.2	923.2
Total CH_4 + NOx (agricultural sector emission)	447.5	447.9	454.1
Percentage of total CH_4 + NOx emission (%)	45%	47%	49%

In fact, prior to the 1900s land use was the dominant global source of CO_2 from human activity, and since 1850 an estimated 160 billion metric tons of carbon from biomass and soils have been emitted worldwide as a consequence of land use and land-use changes [45]. Agricultural soils produce the majority of NO_x emissions in the United States. Estimated emissions from this source in 2006 were 265.0 Tg CO_2 Eq. Direct increases occur through a variety of management practices that add, or lead to greater release of, mineral N to the soil, including fertilization; application of managed livestock manure and other organic materials such as sewage sludge; deposition of manure on soils by domesticated animals in pastures, rangelands, and paddocks (PRP) (i.e., by grazing animals and other animals whose manure is not managed); production of N-fixing crops and forages; retention of crop residues; and drainage and cultivation of organic cropland soils (i.e., soils with a high organic matter content, otherwise known as histosols). Other agricultural soil management activities, including irrigation, drainage, tillage practices, and fallowing of land, can influence N mineralization in soils and thereby affect direct emissions. Mineral N is also made available in soils through decomposition of soil organic matter and plant litter, as well as asymbiotic fixation of N from the atmosphere [104].

Table 11. Direct N emissions from agricultural soils by land use (Tg CO_2 Eq.) Source: [104]

Activity	**1990**	**2000**	**2006**
Cropland	130.9	142.0	138.9
Mineral Soils	*128.1*	*139.1*	*136.1*
Synthetic fertilizer	51.3	55.8	53.6
Organic Amendments (a)	9.4	10.2	10.7
Residue N (b)	9.0	10.2	10.1
Other (c)	58.4	62.8	61.7
Organic Soils	*2.8*	*2.9*	*2.9*
Grassland	87.4	74.0	75.8
Total	218.3	216.0	214.7

(a) Organic amendment inputs include managed manure amendments and other commercial organic fertilizer (i.e., dried blood, dried manure, tankage, compost, and other).

(b) Residue N inputs include unharvested fixed N from legumes as well as crop residue N.

(c) Other N inputs include mineralization from decomposition of soil organic matter as well as asymbiotic fixation of N from the atmosphere. The key to reducing net GHG emissions from agriculture is to promote practices that maintain or increase carbon stocks, while at the same time increasing the efficiency of agricultural inputs (e.g., fertilizer, irrigation, pesticides, animal feed, and animal waste). GHG mitigation policies should address both carbon stocks and emissions of nitrous oxide and methane to achieve the best overall mitigation results [73].

Chapter 13

IMPACT OF THE KYOTO PROTOCOL ON AGRICULTURE

The Kyoto Protocol to the United Nations Framework Convention on Climate Change is a proposed treaty that would require industrialized countries to reduce greenhouse gas emissions by sharply restricting their use of coal, oil, and natural gas. Because U.S. agriculture accounts for nearly one-fifth of U.S. greenhouse gas emissions, compliance with the Kyoto Protocol could increase U.S. farm production expenses by 10 billion dollars to 20 billion dollars annually and depress annual farm income by 24 percent to 48 percent. Higher fuel oil, motor oil, fertilizer, and other farm operating costs would also mean higher consumer food prices, greater demand for public assistance with higher costs, a decline in agricultural exports, and a wave of farm consolidations. In short, the Kyoto Protocol represents the single biggest public policy threat to the agricultural community today [27].

U.S. farming is very energy intensive. Fuel and oil costs represent only about 30 percent of the total energy bill farmers pay. The remaining 70 percent is hidden in the prices of manufactured inputs, fertilizer, pesticides, and other chemicals. Corn and cotton crops use more pesticides and nitrogen fertilizer, products that are quite sensitive to changes in energy prices. Higher energy taxes have the potential for causing an economic downturn in the agricultural sector [27].

Chapter 14

GLOBAL WARMING

Rising temperatures due to global warming will likely stimulate soil organic matter decomposition, which may reduce or eliminate the potential to further increase soil carbon stocks [73].

Between 1945 and 1986, the amount of carbon released from UK land tripled. Since 1980, UK soils lost 12-15% carbon of the total carbon in the soil, due mainly to loss from arable lands. This is about four million tonnes/yr lost in greenhouse gas emissions. There has been a reduction in GHG emissions from UK agriculture since 1990 of about 14%. A study in the journal *Nature* looked at the carbon content of soil in England and Wales from 1978-2003 and found that it fell steadily, with some 13 million tonnes of carbon released from British soil each year. Much of it may be entering the atmosphere as the greenhouse gases carbon dioxide and methane, which scientists say have caused global warming. Since the carbon appeared to be released from soil regardless of how the soil was used, they concluded that the main cause must be climate change itself [8].

Soil carbon content depends on rates of addition from plant growth versus rates of removal in decomposition, leaching and other soil processes, and each of these is sensitive to changes in land use, climate and other variables. The fact that the losses appear to be happening across countries irrespective of land use suggests a link to climate change. Over the survey period, the mean temperature across England and Wales increased by about 0.58°C and there were also changes in rainfall distribution. Climate change will affect soil carbon turnover through various processes. Increases in temperature will tend to increase rates of organic matter decomposition by soil microbes, although the magnitude of this effect and differences between soils are uncertain. The effects of temperature will interact in a complicated way with changes in soil moisture brought about by changing rainfall and evapo-transpiration patterns, and changes in atmospheric CO_2 and nitrogen deposition. In freely drained soils, warmer, drier conditions may retard decomposition of organic matter if lack of moisture limits soil microbes. But in wet anoxic soils, increased oxygenation of the soil as evaporation increases the depth to the water table will greatly reinforce the increase in decomposition with increased temperature. Such effects may in part explain the increase in the relative rate of loss with soil carbon content because more organic soils tend to be wetter. Also consistent with this, more organic soils necessarily contain a greater proportion of slowly decaying organic matter, and the rate of turnover of this material appears to be more sensitive to temperature changes than that of more labile organic matter. However, more labile, the magnitudes of these effects are unknown[8].

Chapter 15

IMPORTANCE OF SOM AND EFFECTS OF LOSSES OF SOIL CARBON

More than twice as much carbon is held in soils as in vegetation or in the atmosphere, and changes in soil carbon content can have a large effect on the global carbon budget [8]. Soil deterioration from depletion of organic matter is an increasingly serious global problem that contributes to hunger and malnutrition. Often a result of unsustainable farming, overuse of chemical fertilizers and drought, the main weapons to combat the problem — compost, animal manure and crop debris — decompose rapidly [4]. Earth's soil is the largest terrestrial pool of carbon. In other words, most of the earth's carbon is fixed in soil. But if this soil is intensively cultivated by tillage and chemical fertilization, organic matter in soil will be quickly decomposed into carbon dioxide by soil microbes and released into the atmosphere, leaving the soil compacted and nutrient-poor [4]. Most soil fertility experts believe that higher amounts of soil organic matter are related to increased productivity because of its contribution to water holding capacity, improved soil structure, and supply of nutrients [42]. The lack of good rotations on most crop farms, partially caused by the availability of inexpensive synthetic fertilizers, has resulted in a loss of soil organic matter and a decrease in the diversity of organisms in the soil. This degradation of soil quality allows the growth of large populations of disease organisms and plant parasites that would have been held in check by a diverse community of competing organisms. Also, plants that are unhealthy tend to attract more insect pests than healthier plants.

SOM (soil organic material) is defined here as the nonliving component of organic matter in soil. The ultimate source of organic matter in soils is CO_2 fixed by plants, including leaf litter, roots, and root exudates. The activity of soil fauna (especially fungi and microbial communities) metabolizes some of these substrates and transforms others into more resistant organic compounds (collectively referred to as humus). The stabilization and fate of organic matter residues is affected by the quality of the original plant substrate and the physical environment in the soil (clay content and mineralogy, pH, O_2 availability), formation and disruption of soil aggregates. Carbon is lost from soil mostly as CO_2 produced during decomposition of organic matter, though losses of carbon through leaching or erosion may be important when considering C balance in soils on long time scales [99].

Organic matter in soil plays important roles in determining soil water-holding capacity and soil structure, and provides a long-term store of nutrients needed by plants. Thus, changes in SOM will have important feedbacks on hydrology and plant productivity [99].

There is general agreement that SOM contains at least three identifiable C pools: root exudates and rapidly-decomposed components of fresh plant litter (''active'' pool), stabilized organic matter that persists in soils over several thousands of years (''passive'' pool), and a poorly-defined ''intermediate'' or ''slow'' C pool that has turnover times in the range of years to centuries [99].

The mechanisms through which climate or land-use change may influence soil carbon cycling include a combination of biotic controls that affect the amount and resistance to decay of C added to soils by plants; physical controls, including the area and chemistry of surfaces for stabilization of organic matter; and the availability of oxidants to decomposers. These mechanisms operate on vastly different time scales [99]. Soil minerals, forming over millennial time scales, control how much of the soil carbon is chemically and physically protected from decomposition. Finally, ecosystem disturbance caused by fire, flooding, tree blow-down, etc. will cause drastic changes in plant production and soil conditions for decomposition that far exceed those associated with interrannual variability [99].

After CO_2 fertilization, the most often-cited mechanism leading to a terrestrial carbon sink is enhanced nitrogen availability to plants due to enhanced nitrate deposition. Nitrogen availability also will be increased if there is net decomposition of SOM, because SOM is a major source of nitrogen for plants. Although nitrogen mineralization is accompanied by CO_2 efflux, the transfer of nitrogen from soil to plants would result in net carbon sequestration because the C:N ratio of plants is roughly 10 times the ratio in SOM [99].

The amount of organic matter in soils that is positively correlated with moisture and negatively correlated with precipitation and temperature is a major control of the turnover rate of fast-cycling soil C. Both vegetation inputs and decomposition rates are expected to increase with increases in temperature. Soils become an amplifying feedback if the temperature dependence of decomposition is steeper than that of plant productivity [99].

One consequence of all studies of C turnover as a function of temperature is that soils already should be a net source of C to the atmosphere due to increased decomposition rates because of documented temperature increases over the past century. This present net C loss presumably would offset hypothetical increases in soil C stores from increased plant production due to CO_2 fertilization [99]. Climate, vegetation, parent material, and time all affect the processes controlling accumulation and decomposition of organic matter in soils, as has been known for decades [99]. Soil organic carbon inventories increase regionally from arid to wet environments, whereas incubation and field decomposition studies show increases in decomposition rates with added moisture in aerobic soils. Decomposition rates for organic matter decrease dramatically under waterlogged conditions. Low C inventories in arid soils may be due to decreased C inputs, faster C turnover, or differences in the quality of plant material. As yet, the relative contributions of these factors are unknown [99].

Soil organic matter consists of a variety of components. In varying proportions and many intermediate stages, these include the following:

- raw plant residues and microorganisms (one to 10 percent)
- active organic fraction (10 to 40 percent)
- resistant or stable organic matter (40 to 60 percent), also referred to as humus [61]

Soil organic matter has the following affects on soil quality [17]:

- stores and supplies plant nutrients (N.P.K. and micronutrients, increases cation exchange capacity, and mitigates mineral deficiency stress.
- stabilizes and holds soil particles together as aggregates.
- helps soil to resist compaction, promotes water infiltration, and reduces runoff.
- aids growth of crops by improving the soil's ability to store and transmit air and water, (as measured by improved porosity), water holding capacity, and drought resistance.
- makes soil more friable and easier to work so that plant roots can penetrate the soil profile better.
- provides a source of carbon and energy for soil microbes that cycle nutrients and fight plant diseases. A soil rich in organic matter and regularly supplied with different kinds of soil organic matter will support a rich and varied population of soil organisms. Organic matter provides a carbon source for primary producers like cyanobacteria that can convert atmospheric nitrogen to plant available N forms. Organic matter is the principle food source for secondary consumers. A soil populated by a diverse, active microbial population is less likely to support uncontrolled spread of plant pathogens. Enhanced microbial activity also accelerates the breakdown of pesticides.
- reduces the negative environmental effects of pesticides, heavy metals and other pollutants by binding contaminants; binds organic pollutants, keeping them out of the soil solution where they would be taken up by plants or leached into ground water. Organic matter also provides sites for microbes to colonize and decompose organic pollutants.
- has been shown to protect plant roots from potentially harmful levels of aluminum, which occur in some acid soils, and also from certain herbicides that are potentially toxic to plants.
- soil pH buffering capacity: organic matter has the ability to moderate major changes in the soil pH. Organic matter buffers the soil against major swings in pH by either taking up or releasing H^+ into the soil solution, making the concentration of soil solution H^+ more constant. The result is a stable pH close to neutral or suitable for the specific crop to be grown.

Soil organic matter also binds plant micronutrients like iron, aluminum, zinc, copper and manganese by chelation—a chemical association of organic matter and micronutrients that keeps the micronutrients in a form available for plant uptake [17].

When fresh organic matter is added to the soil, soil microbes release long-chain sugars or polysaccharides relatively quickly. These polysaccharides promote formation of large or macro-aggregates. As the organic matter decomposes over the longer term, different sizes of aggregates are formed that are resistant to physical disruption. The number and diversity of stable soil aggregates are what give a soil an excellent physical structure [17].

A soil with a good physical structure will have pores and channels of many sizes, from the tiniest channel to large spaces that will allow rainfall to penetrate without running off and taking soil with it. Soil organic matter helps to form and maintain the air passages and channels, protecting the soil from compaction [17].

The resistant or stable fraction of soil organic matter contributes mainly to nutrient holding capacity (cation exchange capacity) and soil color. This fraction of organic matter decomposes very slowly and, therefore, has less influence on soil fertility than the active organic fraction [61]. Organic matter in soil serves several functions. From a practical agricultural standpoint, it is important for two main reasons: first as a revolving "nutrient bank account," and second, as an agent to improve soil structure, maintain tilth and minimize erosion. The stable organic fraction (humus) adsorbs and holds nutrients in a plant-available form. Organic matter does not add any new plant nutrients, but it does release nutrients in a plant-available form through the process of decomposition. If the rate of addition is less than the rate of decomposition, soil organic matter will decline; and, conversely, if the rate of addition is greater than the rate of decomposition, soil organic matter will increase. The term steady state has been used to describe a condition whereby the rate of addition is equal to the rate of decomposition [61].

Before our soils were cultivated, they had achieved a steady state. In most of our prairie soils, the increased rate of decomposition associated with cultivation, combined with the low rates of crop residue addition associated with crop-fallow rotations, has caused a fairly rapid decline in soil organic matter. The rate of decline decreases with time as the amount of total soil organic matter decreases and, particularly, as the active organic fraction is depleted. Cultivation of soils that are naturally high in organic matter will usually result in a decrease of organic matter [61].

Agricultural soils contain substantial amounts of carbon, typically 20 to 80 tonnes per hectare in the top 20 cm. However, relative to their native ecosystem levels, most agricultural soils are depleted in carbon, having lost 30-50% of their original carbon levels due to changes associated with production agriculture and past management practices.

Historically, agricultural practices often resulted in reduced inputs of carbon through plant residues and increased losses via decomposition and erosion. Lower productivity, particularly prior to the 1950s, and greater removal of crop residues decreased the amount of plant material that could potentially add carbon to the soil [11]. Agricultural soils typically contain one to four percent C on mass basis. The relatively small amounts of C could be considered analogous to a catalyst or to the tip of the iceberg where a small amount visible has a big impact. [77] Carbon loss can also affect soil productivity, soil quality, nutrient cycling, soil fertility, and, ultimately, crop production. In agricultural production systems, C loss may occur as a result of biomass burning, intensive tillage or cultivation, or crop residue removal [87].

Consistent evidence of an organic carbon decline for fertilized soils throughout the world had been found [52].

Chapter 16

ENVIRONMENTAL BENEFITS OF SOIL ORGANIC CARBON

Soil organic matter is so valuable for what it does in the soil that it can be referred to as "black gold" because of its vital role in physical, chemical and biological properties and processes within the soil system [77]. Storing C will also lead to enhanced air quality, water quality and increased productivity as well as mitigating the greenhouse effect [77]. Carbon sequestration means to "capture and secure or store", in this case soil C. [77] The main goal is to enhance C fixation through photosynthesis and to reduce soil C emissions by securing and storing C that might otherwise be lost to the atmosphere. The soil contains two to three times as much C as the atmosphere. By minimizing the increase in ambient CO_2 concentration through soil C management, we minimize the production of greenhouse gases and minimize potential for climate change [77].

The mineral P fertilizer and equivalent amounts of manure did not result in higher above- or below-ground biomass and nutrition in comparison to the unfertilized control. Cattle manure has a similar effect to chicken manure on plant growth and nutrition. In contrast, a significant growth improvement was found for the Anthrosol and after charcoal amendments in ferralsol. Nitrogen uptake of cowpea was significantly ($P < 0.05$) decreased by charcoal additions and in anthrosol, which was an effect of poor N nutrition. Phosphorus nutrition and uptake increased when charcoal was added to ferralsol and for the anthrosol. Charcoal amendments improved foliar K nutrition and uptake in contrast to the anthrosol, whereas K nutrition was even significantly ($P < 0.05$) reduced in comparison to the control. Increasing the amount of charcoal further increased above- and belowground biomass production. Similarly, P, K, Ca, Zn, and Cu uptake by cowpea increased.[59] The Anthrosol contained twice as much soil C compared to ferralsol. Charcoal amendments to ferralsol, however, resulted in the highest soil C contents which were doubled compared to anthrosol [77].

Enhanced soil C storage and management also helps the environment win as improvements in soil, air and water quality are all enhanced with increased amounts of C within the soil. Increasing soil C storage can increase infiltration, increase fertility, decrease wind and water erosion, minimize compaction, enhance water quality, decrease C emissions, impede pesticide movement and enhance environmental quality. Accepting the challenges of maintaining food security by incorporating C storage in conservation planning demonstrates concern for our global resources. This concern presents a positive role for conservation agriculture that will have a major impact on global sustainability and our future quality of life

[77]. Not only the nutrient contents but also the nutrient retention can be improved with charcoal additions to soil, but not with the addition of coal degradation products such as coal ash or fly ash. This is especially important in highly weathered soils with low ion-retention capacities. Adding charcoal to soil can significantly increase seed germination, plant growth, and crop yields. Low SOM contents may be responsible for the low available water capacity and the weak structure of many agricultural soils. Charcoal may not only change soil chemical properties, but also affect soil physical properties such as soil water retention and aggregation. These effects may enhance the water availability to crops and decrease erosion [32].

The use of charcoal as a soil conditioner for sustainable agriculture in the humid tropics presents the following advantages:

High nutrient content and nutrient retention capacity lead to improved nutrient supply for plants and reduced nutrient losses by leaching. We assume that two processes are responsible for this: (1) nutrients are physically trapped in the fine pores of amorphous carbonized materials, and (2) slow biological oxidation produces carboxylic units on the edges of the condensed aromatic backbone of the charcoal,which increases the CEC.

Transformation of labile plant organic matter into stable C pools can reduce the release of the greenhouse gas CO_2 into the atmosphere during land clearing and can increase C sequestration in the soil. There are strong indications that charcoal is very slowly mineralized in the soil environment. Charcoal from fallow vegetation and/or organic wastes can be easily produced by local farmers and also by those with low income. Charcoal production is a well-known technique and the required tools and resources are readily available [32].

Interactions of bio-char with soil microorganisms are complex. On one hand, soil microbial diversity and population size, as well as population composition and activity, may be affected by the amount and type of bio-char present or added to soil. On the other hand, microorganisms are able to change the amount and properties of bio-char in soil. Both effects have significant influence on nutrient cycles and nutrient availability to plants. A higher retention of microorganisms in bio-char soils may be responsible for greater activity and diversity due to a high surface area as well as surface hydrophobicity of both the microorganisms and bio-char. Activated carbon, which is chemically similar to black carbon or bio-char in soil, has been shown to sorb microorganisms strongly, and this adsorption has been seen to increase with higher hydrophobicity. Bio-char additions not only affect microbial populations and activity in soil, but also plant–microbe interactions through their effects on nutrient availability and modification of habitat [56].

Bio-char amendments have so far shown beneficial effects when applied to rice, sorghum, corn, various beans (soybean, common bean, cow pea, moongbean), banana, and vegetables such as carrots. Since bio-char is a valuable commodity, applications can be successfully applied to smaller units of area than entire fields, such as high-fertility trenches in contour rows of steep land for vegetables or grain crops, and planting holes for tree crops [56].

The most common system responses attributed to microfloral grazers (protozoa, nematodes, microarthropods) in the literature are increased plant growth, increased N uptake by plants, decreased or increased bacterial populations, increased CO_2 evolution, increased N and P mineralization, and increased substrate utilization. Based on this evidence in the literature, a conceptual model was proposed in which microfloral grazers were considered as separate state variables. To help evaluate the model, the effects of microbivorous nematodes on microbial growth, nutrient cycling, plant growth, and nutrient uptake were examined with reference to activities within and outside of the rhizosphere. Blue grama grass (*Bouteloua gracilis*) was grown in gnotobiotic microcosms containing sandy loam soil low in inorganic N, with or without chitin amendments as a source of organic N. The soil was inoculated with bacteria (*Pseudomonas paucimobilis* or *P. stutzeri*) or fungus (*Fusarium oxysporum*), with half the bacterial microcosms inoculated with bacterial-feeding nematodes (*Pelodera* sp. or *Acrobeloides* sp.) and half the fungal microcosms inoculated with fungal—feeding nematodes (*Aphelenchus avenae*). Similar results were obtained from both the unamended and the chitin—amended experiments. Bacteria, fungi, and both trophic groups of nematodes were more abundant in the rhizosphere than in nonrhizosphere soil. All treatments containing nematodes and bacteria had higher bacterial densities than similar treatments without nematodes. Plants growing in soil with bacteria and bacterial-feeding nematodes grew faster and initially took up more N than plants in soil with only bacteria, because of increased N mineralization by bacteria, NH_4^+—N excretion by nematodes, and greater initial exploitation of soil by plant roots. Addition of fungal-feeding nematodes did not increase plant growth or N uptake because these nematodes excreted less NH_4^+—N than did bacterial-feeding nematode populations and because the N mineralized by the fungus alone was sufficient for plant growth. Total shoot P was significantly greater in treatments with fungus or *Pelodera* sp. than in the sterile plant control or treatments with plants plus *Pseudomonas stutzeri* until the end of the experiment. The additional mineralization that occurs due to the activities of microbial grazers may be significant for increasing plant growth only when mineralization by microflora alone is insufficient to meet the plants' requirements. However, while the advantage of increased N mineralization by microbial grazers may be short-term, it may occur in many ecosystems in those short periods of ideal conditions when plant growth can occur. Thus, these results support other claims in the literature that microbial grazers may perform important regulatory functions at critical times in the growth of plants [48].

Boreal Forest Wild Fire Investigations

Boreal forest wild fire experiments were conducted in each of three contrasting boreal forest sites in northern Sweden [112]. Mesh bags were filled with pure humus collected from the forest, pure charcoal created in the laboratory, or a 50:50 mixture of humus plus charcoal. These bags were left in the field and harvested over 10 years. Boreal forest wild fire investigation cases found that, over the 10-year period, loss of mass and C from the bags containing mixtures of charcoal and humus was substantially greater than what was expected on the basis of the components considered separately. Further, nitrogen immobilization was less than expected in the mixture bags. Given that charcoal decomposition rates in soil are extremely slow and that in our study system charcoal persists for thousands of years in the humus layer without evidence of mass loss, most of the enhanced loss of mass and C caused by mixing charcoal and humus must have resulted from charcoal promoting humus loss rather than humus promoting charcoal loss. These results are consistent with charcoal particles serving as foci for adsorption of organic compounds and microbial growth and activity, leading to enhanced decomposition rates and mass loss of associated humus, again in boreal forest cases after wild fire. These Swedish findings indicate that charcoal can promote rapid loss of forest humus and below ground C during the first decade after its formation. Fire often causes substantial losses of ecosystem C, and our results provide evidence for a previously unreported mechanism that could contribute to these losses. Results show that these effects can be partially offset by its capacity to stimulate loss of native soil C in boreal forests. The effect of charcoal on native soil C needs to be explicitly considered to better understand the potential of black C as an ecosystem C sink and agent of C sequestration [112]. Forest wild fire conditions and effects need to be further investigated.

Chapter 17

EFFECTS OF BIOCHAR APPLICATION

TERRA PRETA

Fifteen hundred years ago, tribes people from the central Amazon basin mixed their soil with charcoal derived from animal bone and tree bark. Today, at the site of this charcoal deposit, scientists have found some of the richest, most fertile soil in the world. Now this ancient, remarkably simple farming technique seems far ahead of the curve, holding promise as a carbon-negative strategy to rein in world hunger as well as greenhouse gases [4]. Tropical forests account for between 20 and 25% of the world terrestrial carbon (C). Soils under tropical forest contain approximately the same amount of C as the lush vegetation above it. The current conversion of Amazonian forest to agricultural land makes disturbance of this C stock important to the global C balance and net greenhouse gas emissions. Changes in land use, particularly by clearing forests, reduce organic C by 20% to 50% in the upper soil layers [92]. Furthermore, this reduction of soil organic matter (SOM) is causing soil degradation. Thus, agriculture is not sustainable without nutrient inputs beyond three years of cultivation. The efficiency of conventional fertilizers (such as nitrogen N) is limited by a low nutrient retention capacity conjoined with strong tropical rains. On the other hand, large amounts of phosphate fertilizers are needed to overcome the soil'.s high P-fixation capacity [92]. To overcome these limitations, slash-and-burn agriculture (shifting cultivation) is practiced by about 300 to 500 million people, affecting almost one third of the planet's 1500 million ha of arable land [92]. The existence of an anthropogenic and C-enriched dark soil in different parts of the world, and especially in Amazonia (Amazonian dark earths ADE or Terra Preta de Índio), proves that the predominant ferralsols and acrisols can be transformed into fertile soils. The ADE.s fertility is most likely linked to an anthropogenic accumulation of phosphorus (P), calcium (Ca), and black C as charcoal. Charcoal persists in the environment over centuries and is responsible for the stability of the ADE's SOM [92]. Charcoal formation and deposition in soils seems to be a promising option to transfer an easily decomposable biomass into refractory SOM pools [92].

Upland soils in the humid tropics such as the Amazon basin are often highly weathered and, therefore, possess low plant-available nutrient contents. This is a result of both high rainfall and low nutrient retention capacity. In these soils, the cation exchange capacity is very low due to the dominance of iron and aluminum oxides and kaolinite, as well as generally low soil organic matter contents and pH values [59]. Two basic approaches can be used to reduce nutrient leaching: applying slow-release nutrient forms such as organic fertilizers; and increasing adsorption sites, thereby retaining applied inorganic nutrients. Additionally, organic fertilizers may mineralize rapidly and, therefore, have to be applied repeatedly to maintain sufficient available nutrients and exchange capacity. Applications of organic matter may be a feasible way to create sustainably fertile soils similar to the prehistoric anthrosols. An additional option to manure and composts could be the application of charcoal to soil [59].

When charcoal was applied in un-weathered condition, the microbiological parameters (respiration, biomass, population growth, and efficiency) increased linearly and significantly with increasing charcoal concentrations (50, 100 and 150 g kg-1 soil). Application of pyrogenous acid caused a sharp increase in soil respiration, biomass, and reproduction, possibly because the condensates from smoke contain easily-degradable substances which could be utilized by the microbes for their metabolism [92]. In a greenhouse experiment, leaching of N was significantly ($P < 0.05$) reduced if ammonium sulphate was applied with charcoal. Charcoal simultaneously served as K fertilizer and increased the retention of N. Long-lasting soil fertility improvement due to organic fertilization and a synergistic effect if both charcoal and mineral fertilizer were applied was observed in a field experiment [92].

Mineral fertilized soils amended with charcoal and terra preta soils had a significantly higher potential for microbial population growth coupled with a low microbial respiration in absence of an easily degradable C source (glucose). The soil respiration before substrate addition correlated positively with the population growth rate on the plots, whereas terra preta had a very low soil respiration and very high population growth after substrate additions. Forest soils had a higher respiration rate but a very low population growth. These results reflect the relatively high biodegradable SOM content of primary forest topsoil but low available nutrients (requirement for microbial population growth), in contrast to refractory terra preta SOM with high available soil nutrient contents [92].

The N recovery in biomass was significantly higher on compost amended plots due to significantly higher biomass production. The retention in soil was significantly higher in the charcoal-amended plots after the second harvest due to higher N retention and cycling of crop residues which remained on the plots after harvesting. Total N recovery (in soil, crop residues and grains) was significantly ($P < 0.05$) higher on charcoal (18.1%), charcoal plus compost (17.4%), and compost (16.5%) treatments in comparison to only mineral-fertilized plots (10.9%) [92]. Charcoal influences soil quality in manifold ways, most importantly by reducing available Al and reducing acidity. Furthermore, charcoal adds K to the soil and has the potential to reduce N leaching. Charcoal amendments increased the reproduction rate of the microbial population after substrate addition whether or not the plot was fertilized. The effects of charcoal on soil biological, chemical and physical properties are complex, making it difficult to isolate single significant charcoal effects but, added up, they caused significantly increased plant growth and crop production [92].

At the 235th national meeting of the American Chemical Society, scientists reported that charcoal derived from heated biomass has an unprecedented ability to improve the fertility of soil — one that surpasses compost, animal manure, and other well-known soil conditioners [4].

They also suggested that this so-called "biochar" profoundly enhances the natural carbon seizing ability of soil. Dubbed "black gold agriculture," scientists say this "revolutionary" farming technique can provide a cheap, straight-forward strategy to reduce greenhouse gases by trapping them in charcoal-laced soil [4].

Soils receiving charcoal produced from organic wastes were much looser, absorbed significantly more water and nutrients and produced higher crop biomass. The authors, with Delaware State University, say "the results demonstrate that charcoal amendment is a revolutionary approach for long-term soil quality improvement"[4].

Applying raw organic materials to soil only provides a temporary solution, since the applied organic matter decomposes quickly. Converting this unutilized raw material into biochar, a non-toxic and stable fertilizer, could keep carbon in the soil and out of the atmosphere [4].

With its porous structure and high nutrient- and water-holding capabilities, biochar could become an extremely attractive option for commercial farmers and home gardeners looking for long-term soil improvement. The biochar-infused soil showed vastly improved germination and growing rates compared to regular soil [4].

Many soils of the lowland humid tropics are thought to be too infertile to support sustainable agriculture. However, there is strong evidence that permanent or semi-permanent agriculture can itself create sustainably fertile soils known as 'terra preta' soils. These soils not only contain higher concentrations of nutrients such as nitrogen, phosphorus, potassium and calcium, but also greater amounts of stable soil organic matter. Frequent findings of charcoal and highly aromatic humic substances suggest that residues of incomplete combustion of organic material (black carbon) are a key factor in the persistence of soil *organic matter in these soils*. Our investigations showed that 'terra preta' soils contain up to 70 times more black carbon than the surrounding soils. Due to its polycyclic aromatic structure, black carbon is chemically and microbially stable and persists in the environment over centuries. Oxidation during this time produces carboxylic groups on the edges of the aromatic backbone, which increases its nutrient-holding capacity. The author concludes that black carbon can act as a significant carbon sink and is a key factor for sustainable and fertile soils, especially in the humid tropics [31].

Soil Fertility

There is a long tradition in Japan of using charcoal as a soil improver. The idea that the application of charcoal stimulates indigenous arbuscular mycorrhizal fungi in soil and thus promotes plant growth is relatively well-known in Japan, although the actual application of charcoal is limited due to its high cost [67].

Soil Water Retention

Hydrophobicity of soils can have severe adverse effects, such as reduced plant growth and increased overland flow leading to increased soil erosion. It is, therefore, important to consider the possibility that biochar applications may introduce hydrophobic compounds into the soil [20].

Water repellency of soils is sometimes observed to increase after fires. In part, this may be due to physical changes in the soil whereby small particles of ash and char block soil pores and reduce water infiltration rates. Of more concern in the context of biochar is the mechanism whereby hydrophobic organic compounds are produced during combustion and coat soil particles. Organic coatings are a common cause of water repellency in soils. The possibility of such organic compounds or tar residuals occurring in low grade and for energy application produced biochar therefore, there is a cause for appropriate soil application and ecotoxicology concern [115].

The processes involved in distributing volatile organic compounds (VOCs) within the soil will be different in the case of biochar application compared to fires occurring directly on the soil [115].

In the first instance, when char is artificially made, the VOCs will be adhered to the char internal particles and external surface only, so this low grade char may act as biotoxic and water-repellent centres. In the second case, the heat of the wild fire will distribute the VOCs throughout the soil structure, while thermally impacting topsoil organic and inorganic constituents — allowing it also to coat non-char soil particles. Thus, even if the same hydrophobic compounds were present in both cases, we may expect the effect on soil water repellency effects to be different. Nonetheless, the introduction of recalcitrant hydrophobic compounds into the soil with biochar would leave open the possibility of their subsequent redistribution within the soil by physical, chemical or biological processes [115].

Under any circumstances organic tars are biotoxic compounds, occurrence of which depends on pyrolysis conditions, applied technology performance and feedstock characteristics. Biochar organic tar residuals, therefore, are to be avoided, especially for agricultural soil applications, and its content should be maximized to 1% weight by weight of the biochar [115]. Notwithstanding the possibility of introducing hydrophobic compounds to the soil by biochar application, there is some evidence suggesting that, at least in some instances, biochar may have the opposite effect of increasing soil water retention [115]. It was found that charcoal increased the available moisture in sandy soil, had no effect in loamy soil, and decreased the available moisture in clayey soil. The increase in available moisture observed in sandy soils may make biochar a useful tool in the reversal of desertification [115]. Charcoal, in many soils and sediments, is considered a high-affinity sorbent for organic pollutants [81].

Reduction of Nutrient Leaching

It has been suggested that biochar may have the potential to reduce leaching of pollutants from agricultural soils. This possibility is suggested by the strong adsorption affinity of biochar for soluble nutrients such as ammonium, nitrate, phosphate, and other ionic solutes. If this affinity of biochar for ionic solutes can in fact be utilised to reduce run-off in agricultural watersheds, it will have important benefits in terms of reducing hypoxia of inland and coastal waterways caused by eutrophication. [57, 31, 75].

Chapter 18

How Much Biochar Can Be Added to Soil?

The amount of biochar that can be added to soil before it ceases to function as a beneficial soil amendment and becomes detrimental will be the limiting factor in the use of biochar as a soil additive. The strongest evidence that high concentrations of black carbon in soil may be beneficial under some conditions comes from the Amazonian dark earths (ADEs) such as terra preta and terra mulata – charcoal-rich soils which contain approximately three times more soil organic matter, nitrogen and phosphorus than adjacent soils and have twice the productivity [31].

Total amounts of black carbon found in terra preta soils were 25±10 Mg ha-1 and 25±9 Mg ha–1 at 0–30 cm and 30–100 cm soil depths, respectively [31] .

It is important to note, however, that terra preta data come principally from studies on highly weathered tropical soils with very low natural SOC levels. Much less is known about the effect of standard biochar additions to relatively fertile temperate soils. The lack of research on such soils arises because the potential benefits of raised fertility are unlikely to be as great as in regions with soils of low natural fertility [115].

Versus the terra preta tropical applications, the 3R-AGROCARBON product composite dose has been high efficient at temperate soil horticultural farming at doses as low as 400 kg/ha to 1250 kg/ha.

Chapter 19

Enchanced Soil Microbiological Activity by Charcoal

It has been found that "apparently, some microorganisms were able to live with BC as sole C source". It was also found that BC in soils may enhance the rate of decomposition of labile C compounds [37].

In laboratory conditions, glucose mineralisation was enhanced over the control by the presence of charred material, i.e., there was an interactive priming of BC and glucose mineralisation. Researchers concluded that BC in soils may promote growth of microorganisms and the decomposition of labile C compounds rather than stabilise them against degradation [37]. Large accumulations of BC in some soils and BC resistance to chemical oxidation suggest that black carbon is stable over at least such a timescale. Nonetheless, considerable uncertainties remain about just how fast biochar may decompose under different soil conditions [115].

Charcoal seems to assist microbial activity by having a porosity that provides a favourable microhabitat, weak alkalinity and by being a substrate unfavourable for saprophytes. AMF fungi easily extend their extraradical hyphae into charcoal buried in the soil and sporulate in the particles, encouraging effects of charcoal on arbuscular mycorrhiza l(AMF) colonisation [80].

The mechanism whereby charcoal stimulates the growth of AMF is described by [67]

Charcoal may stimulate the growth of AMF by the following mechanism: Charcoal particles have a large number of continuous pores with a diameter of more than 100μ . They do not contain any organic nutrients, because of the carbonization process. The large pores in the charcoal may offer a new microhabitat to the AMF, which can obtain organic nutrients through mycelia extended from roots. This may enable the AMF to extend their mycelia far out from the roots, thus collecting a larger amount of available phosphate [67].

The role of mycorrhizal fungi is particularly important in carbon cycling, because they pump the carbon compounds out of the root into a massive network of fine fungal filaments in the soil, where it becomes available to other microbes and also to larger soil organisms like worms, mites and insects. In return, the fungus gathers phosphorus from the soil and delivers it to the plant, helping the plant to grow better. The research confirmed that there were many different fungi in the roots of each plant, but revealed, for the first time, which of these fungi were most active.[100].

Arbuscular mycorrhizal fungi, found living on plant roots around the world, appear to be the only producers of glomalin. Wright named glomalin after Glomales, the taxonomic order that arbuscular mycorrhizal fungi belong to. The fungi use carbon from the plant to grow and make glomalin. In return, the fungi's hairlike filaments, called hyphae, extend the reach of plant roots. Hyphae function as pipes to funnel more water and nutrients—particularly phosphorus—to the plants [15]. As carbon gets assigned a dollar value in a carbon commodity market, it may give literal meaning to the expression that good soil is black gold. And glomalin could be viewed as its golden seal [15]. A strong glue, glomalin is produced by a beneficial fungus that grows on plant roots. The glue comes off the fungus and is deposited on soil particles. This process leads to build-up and stabilization of aggregates [117].

Higher levels of glomalin give greater water infiltration, more permeability to air, better root development, higher microbial activity, resistance to surface sealing (crusts) and erosion (wind/water). A benefit of glomalin is increased aggregate stability which leads to better soil structure which, in turn, leads to better plant production [117]. Soil aggregation is a complex process that is largely dependent upon microorganisms to provide glues that hold soil particles together. These glues are carbon-containing compounds that protect microorganisms from drying out. We are beginning to understand the importance of one group of soil fungi and the glue that is produced in large amounts by these arbuscular mycorrhizal fungi [117].

As a glycoprotein, glomalin stores carbon in both its protein and carbohydrate (glucose or sugar) subunits. Glomalin is causing a complete reexamination of what makes up soil organic matter. That study showed that glomalin accounts for 27 percent of the carbon in soil and is a major component of soil organic matter. Glomalin weighs two to 24 times more than humic acid, a product of decaying plants that up to now was thought to be the main contributor to soil carbon. But humic acid contributes only about eight percent of the carbon. Another team recently used carbon dating to estimate that glomalin lasts seven to 42 years, depending on conditions. It is glomalin that gives soil its tilth—a subtle texture that enables experienced farmers and gardeners to judge great soil by feeling the smooth granules as they flow through their fingers [15]. As a plant grows, the fungi move down the root and form new hyphae to colonize the growing roots. When hyphae higher up on the roots stop transporting nutrients, their protective glomalin sloughs off into the soil. There it attaches to particles of minerals (sand, silt, and clay) and organic matter, forming clumps. This type of soil structure is stable enough to resist wind and water erosion, but porous enough to let air, water, and roots move through it. It also harbors more beneficial microbes, holds more water, and helps the soil surface resist crusting [15]. These fungi are beneficial to plants because hyphae, hair-like projections of the fungus, explore more soil than plant roots can reach and transport phosphorus and some other nutrients to the plant. In return, plants provide carbon for growth of the fungus [117].

Longer hyphae help plants reach more water and nutrients, which could help plants face drought in a warmer climate. The increase in glomalin production helps soil build defenses against degradation and erosion and boosts its productivity [15].

Chapter 20

CARBON SEQUESTRATION

Although most of the increase in greenhouse gas (GHG) concentrations is due to carbon dioxide (CO_2) emissions from fossil fuels, globally about one-third of the total human-induced warming effect due to GHGs comes from agriculture and land-use change. U.S. agricultural emissions account for approximately eight percent of total U.S. GHG emissions when weighted by their relative contribution to global warming. The agricultural sector has the potential not only to reduce these emissions but also to significantly reduce net U.S. GHG emissions from other sectors. The sector's contribution to achieving GHG reduction goals will depend on economics as well as available technology and the biological and physical capacity of soils to sequester carbon [73].

Every tonne of carbon added to, and stored in, plants or soils removes 3.6 tonnes of CO_2 from the atmosphere [73]. Carbon stocks in agricultural soils are currently increasing by 12 million metric tonnes (MMT) of carbon annually. If farmers widely adopt the best management techniques now available, an estimated 70 to 220 MMT of carbon could be stored in U.S. agricultural soils annually. Together with attainable nitrous oxide and methane reductions, these mitigation options represent five to 14 percent of total U.S. GHG emissions [73].

Increasing the organic matter content of soils (which accompanies soil carbon storage) improves soil quality and fertility, increases water retention, and reduces erosion [73]. Due to past management practices, most agricultural soils have relatively depleted stocks of soil carbon compared with native ecosystems [60]. On farmland, intensive cultivation practices, along with comparatively low productivity, harvesting and removal of residues, and soil erosion depleted the carbon stocks of many agricultural soils by 30 to 50 percent or more compared to their condition under native vegetation. A cycle of soil exhaustion and land abandonment, in part, fueled the westward expansion of agriculture in North America [73]. U.S. agricultural soils have lost, on average, about one-third of the carbon they contained before wide-scale cultivation began in the 1800s. Soil science studies suggest that changes in land use and land management practices could increase the carbon content of crop and grazing land soils by 104-318 million metric tons per year [60].

Increasing the quantity of carbon sequestered—or stored—in soils and biomass is an alternative to reducing emissions of carbon and other greenhouse gases (GHG) in an overall strategy to mitigate global climate change and its negative economic and environmental effects. Conceptual discussions in the climate change literature often acknowledge that a unit of carbon emissions reduction and a unit of carbon sequestration may have very different contributions to net GHG mitigation over time. First, to be equivalent with emissions reductions, carbon sequestration must be maintained for a period equal to the time emitted carbon remains in the atmosphere, which is referred to as "permanence". Second, after undergoing a change from one management practice to another (e.g., from conventional tillage to conservation tillage), terrestrial systems tend to move to new equilibrium carbon levels over time. Terrestrial systems then accumulate additional carbon from activity changes for a finite period, until they reach a new "C-stock equilibrium."

Generally, for a unit of carbon sequestered in soil or biomass in year *t* to have the same climate change mitigation effect as a similar unit of emissions reduction in year *t* (and to be of equal mitigation value), the unit sequestered must remain in the soil or biomass for a period of about 100 years [60].

Organic carbon is maintained in soils through a dynamic process. Plants convert carbon dioxide into tissue during photosynthesis; after a plant dies, a portion of the stored carbon is left behind in the soil by decomposing plant residue and a larger portion is emitted back into the atmosphere. Decomposition rates tend to be proportional to the amount of organic matter in the soil. Hence, over time and under relatively constant environmental and management conditions, rates of carbon additions and emissions tend to equilibrate and the amount of organic carbon in soils stabilizes at a constant, or steady-state, level (i.e., the C-stock equilibrium). If the relationship between additions and losses changes due to a change in soil management, the soil will gradually move to a new C-stock equilibrium [60].

On organic farms, increasing soil organic matter and microbial biomass is a fundamental principle to support agro-ecosystem stability.

Agricultural soils play a key role in C cycling and can act as a source or sink of C and thereby impact greenhouse gas emissions [87]. As TOC declines due to cultivation, the more resistant charcoal fraction increases as a portion of the total C [87].

Chapter 21

BIOCHAR PRODUCTION

COMPETITION BETWEEN BIOCHAR AND BIOMASS ENERGY PRODUCTION

When biochar is added to soil, we are essentially choosing to forgo a renewable energy source – the energy that could be released by the combustion of the char. Thus, even though it may be possible with some feedstocks to obtain some energy co-production along with biochar, this will always be less than the amount of energy that might be obtained by complete energy conversion of the original biomass [115].

However, in the practice and in many cases, the prime objective of the solid fuel-based renewable energy production is the conversion of solid fuels to electric power, such as with the efficient turbine techniques, for which efficient electric power production concentrated energy is required.

Therefore, the question arises as to whether it is more efficient in terms of avoided carbon dioxide emissions to use biomass as a source of energy to displace fossil fuels or whether it would be better to sequester a fraction of the carbon in the biomass as biochar and meet energy demand from other sources [115]. Co-production of biochar and energy is possible [115]. However, there is an economical conflict between maximising energy versus biochar production, unless additional, short-term and attractive economical values are added to the biochar use, such as fertilization, plant growth promotion and other simultaneously achieved multiple benefits. If the additional and short-term economical values are higher than the present energy value of the biochar, then the market will automatically select the higher economic value valorization, e.g. the short-term available higher profit.

From an industrial technology point of view, there is a significant technical difference if the char production primary objective is made for energy or biochar, because the two applications require different product quality requirements, as shown in Table 12.

Fowles (2007) used a simplified high level model to compare these two options by considering just two variables – the energy obtained per unit of CO_2 emission for different fuels, and the percentage of carbon in biomass that can be sequestered by pyrolysis. From this analysis it emerged that in almost all cases the reduction in carbon dioxide emissions is greater if biomass is used to produce biochar for sequestration than if the same biomass had

Table 12. Application-based specific 3R-AGROCARBON char characteristics, application concepts and impacts

Items	Energy Char characteristics	3R-AGROCARBON Bio-char characteristics
Organotoxic residuals	Medium level (5-15% w/w) of organic (tar) residuals, related to the needed ignition characteristics and ignition performance	Very low organic (tar) residuals <1% w/w
Heavy metals	Present and allowable EU/US regulation provides up to maximum limits applied for heavy metal content	Very low total and high ecotox targeted heavy metal content. In practical terms the limit should be far below the present and allowable EU/US regulation limits to make the system long term sustainable.
Halogenic residuals	Present and allowable EU/US regulation limits applied for halogen residual content	Low halogen residual content, far below the present and allowable EU/US regulation limits
Pore size structures	Wide range of micro-, meso- and little	Macroporosity bio-optimized
Pore distribution	As of combustor performance need	Size distribution bio-optimized
Surface	Surface characteristic not relevant	Surface characteristic modified and bio optimized
pH	PH is not relevant	Bio-optimized pH
Climate impacts	Negative direct climate impacts in air	Positive climate impact effects directly and indirectly in air, biosphere, soil and subsurface water.
Soil impacts	Soil impacts not applicable	The application targets restoration of natural balance in a complex ecological system, the soil", with long list of biotic and abiotic elements.
Science and technology	The energy char application management requires understanding and high science disciplines from thermal treatment and environmental sciences	The biochar management requires complex understanding and high science disciplines from thermal treatment, agricultural biotechnology, soil and environmental sciences.
Authority permits	Traditional industrial permitting procedures for thermal units	Combined authority permitting: carbonization industrial production and ecotox tests needed for product applications

been used to produce energy to displace fossil fuels. Perhaps the weakest aspect of Fowles' analysis is that he neglects the possibility of combining CCS with biomass energy. Where this option is economically and technically feasible, it has the potential for a greater reduction in atmospheric CO_2 than biochar production as it may simultaneously sequester a greater proportion of the biomass carbon than biochar, while also offsetting a greater amount of fossil fuel use [115]. For all industrial installations and operations, there are direct and indirect economical and environmental costs, for which life cycles are to be analysed case by case.

Table 13. The economical and environmental costs and impact factors of 3R- AGROCARBON production

Economical costs	**Environmental costs and impacts.**
Equipment manufacturing and installation	Low energy
Plant operations	Total recycling and repeated reuse of all output streams for zero emission solution
Input energy use	High energy efficiency utilization
Input feed material	Regionally available agro, forest, food industrial by-product and waste utilization
Input feed and output product transport	Decentralized installations
Output product application adaptation	Organic and low input farming

How quickly this greenhouse gas payback occurs will depend upon the rate at which the biomass would have released greenhouse gases were it not pyrolysed.

The repeated recycle and reuse of all elements of the feed stream is one of the most important factors for sustainable 3R-AGROCARBON industrialization as a zero-emission solution.

Chapter 22

POTENTIAL AGROCARBON: BIOCHAR FOODSTOCK CHARACTERIZATION

THE CHARACTERISTICS OF THE FOOD AND MEAT INDUSTRIAL ANIMAL BONE-BASED CHARCOALS:

Organic wastes from the human food chain are products directly derived from food industrial processing, which may be fit for human consumption in accordance with community legislation, but are not intended for human consumption for commercial or other reasons. Despite the high qualities of the food chain inputs for human consumption, the technical by-products microbiologically affected during and after the processing (cross-contamination), while there is no economical method to more properly handle these low-value by-products. As a result these by-products will rapidly turn into negatively valued problem substances, especially from the meat industry. The conventional uses of meat and bone meals (MBMs) for animal protein consumption have been banned because of microbiological safety issues. As MBMs may contain harmful food-borne pathogens, direct use as fertilizer in agriculture is not feasible as the conventional thermal treatment pof inactivation at 133°C/20min/3 bars heat/pressure may not be sufficient to guarantee the microbiological sterility of the product. Even if the technical by-product is properly inactivated, the organic waste will still be an optimal nutrient substratum for pathogenic microbes. As the food chain processing environment is an optimal environment for these pathogens as well, the risk of cross contamination is very high. These bacterial infections of public health concern are: *Salmonella*, *Shigella* sp., *Campylobacter*, *Escherichia coli*, *Mycobacterium bovis* and others. Therefore, 850°C material core temperature thermal treatment would be required for safe inactivation (EC 1774/2000). Most of the traditional solutions, because of safety reasons, are not viable anymore, nor the disposal or the incineration. (If high P content material, such as bone meal, is incinerated then the P will be highly corrosive in an oxidative environment, resulting in costly maintenance on the boiler steel components throughout the system).

More than half of the animal by-products are not suitable for normal consumption, because of their unsuitable physical, chemical and microbiological characteristics. More than one million tonnes/year of bone by-products are generated from the slaughtering of bovine animals in the EU countries, and more than seven million tonnes/year total throughout the world. The bone by-products should be considered potentially microbiologically risky materials with more than 70% organic matter content. The technical by-products are safe materials at the point of their genesis; however, if no rapid transformation is implemented, they may turn into risky materials. The total revenue from the sale of animal technical by-products is estimated to be on the order of 1.5 bn € in 2007. The cost of disposing of all animal by-products, largely through incineration, is estimated to be on the order of 3 bn €,and there are also important environmental implications for this disposal route. Therefore, it is important that the farmers do not lose the value of the by-products, but rather have them transformed into added-value product. In practice, there is no financial framework for proper treatment of food waste with conventional technology, as available techniques result in low added value end-products only.

Animal bone charcoals contain high mineral content and low carbon levels. Plant based charcoals contain high carbon content and low mineral levels. In both cases, the carbonization technology performance, thermal conditions and the input material characteristics are the critically important factors for the output charcoal quality performance, especially when applied as biocarbon in relation to a diverse living environment, such as soil. Microorganisms are highly sensitive to ecologically toxic effects from organic and/or inorganic residuals; therefore, advanced carbonization technology is required for production of biospecific chars.

Chapter 23

THE SOLUTION: INTEGRATED 3R-AGROCARBON-TECHNOLOGY FOR RECYCLING AND ADDED VALUE CARBON ECONOMY UPGRADING OF ORGANIC AND INORGANIC AGRO BY-PRODUCTS

The overall objective of the 3R-AGROCARBON technology is to realize integrated thermal inactivation (carbonization) and biotechnological recycling of high P-containing organic waste of non Specified Risk Material (SRM) bone meal and upgrade it into a high added-value and safe biotechnological crop protection and nutrition product for environmentally-friendly vegetable cultivation. The biological control effect targets primarily combat crown rot of tomato and damping off plant pathogens and improvement of plant natural resistance as well. The risk of cross contamination at food chain organic waste streams requires new technological solutions. The developed technology utilizes animal bone char for microbiological carrier and food industrial by-products as nutrients during the solid substrate fermentation and formulation process.

Natural soil-borne microbiological strains are selected for entrapment into the internal sphere of the solid carrier and formulated in a way that long term storage of the substance will be viable in ambient temperature. The method and product provides enhanced survival of the protected microorganism during introduction into the soil.

The 3R Environmental Technology Ltd Group subsidiary, Terra Humana Ltd, has developed an innovative thermal recycling treatment system for high-temperature 850°C carbonization of phosphorous-rich bone meal and integrated to innovative solid substrate fermentation technology. The added-value upgrade of the waste to plant available mineral nutrient product is achieved by integrated thermal and biotechnological means. The production and use of animal bone charcoal mineral as a biotechnological solid carrier is a highly innovative carbonization method and product.

3R-AGROCARBON is a conscious design, maintenance and harmonious integration of agriculturally productive ecosystems which have the diversity, stability, and resilience of natural ecosystems. The soil biological diversity is critically important for variety of life. It reflects the number, variety and variability of living organisms, changing over time, providing diversity within species (genetic diversity), between species (species diversity), and between ecosystems (ecosystem diversity).

Figure 1. Carbonization plant.

The objective driven scientific and technical objectives of the integrated 3R-AGROCARBON recycling and upgrading technology consists of the following main elements:

1) providing a strategy for thermal recycling of phosphorus-rich bone meal and/or other plant-based organic and/or inorganic agricultural by-products into safe and sterile char and mineral, which is suitable as a specific microbiological carrier; the pyrolysis gas-vapor from the 3R carbonization process is reformed to syngas for on site electric power productions.
2) providing objective-driven selection of soil microbial strains compatible with P and other mineral mobilization for plant uptake and biological control effect;
3) including scale-up optimization of innovative SSFF technology;
4) providing design, implementation and tests of an integrated ABC and SSFF "product-like" field demonstration production plant;
5) providing assessment and mathematical descriptions of the 3R-AGROCARBON P dissolution kinetics and the impacts of 3R-AGROCARBON on P and organic matter cycles in soil;
6) providing detailed risk assessment of all components of the input materials, the recycling process and the final product, including the applied microorganisms, in order to establish conclusively that there is no risk to the environment or human health;
7) field demonstrating the 3R-AGROCARBON effectiveness using tomato as the test plant against at least two economically important plant pathogens; field demonstrations are executed under different climatic, soil and agro-industrial

production conditions in Hungary, Israel, Italy, Germany and the Netherlands; the independent and accredited Authority tests are executed in Hungary as of EU permit procedures for plant growth promoter with biological control by-effect;

8) providing justification for socio-economic and technical viability by executed Cost-benefit-analysis, strength-weakness-threats-opportunities and socio-economic assessment of impacts, including the evaluation of its acceptance by consumers and retailers;
9) providing completed industrial scale-up engineering design and a detailed technical implementation plan;
10) providing a clear product and technology-oriented dissemination and marketing strategy, and a plan for increasing the 3R-AGROCARBON technology-based innovation, rapid take-up and commercialization.

In this context, the prevention (of exposure to soil-borne plant diseases by biological control effect), protection (enhancement of plant natural resistance and promotion of plant growth) and preservation (supply of natural, clean and high efficient available plant fertilizer) are important drivers.

The patented 3R carbonization technology is the only known technology which is capable of carbonizing organic material up to 850°C material temperature, the agrocarbon end-product of which is specifically manufactured for biotechnological adaptation purposes, while pyrolysis gas-vapour by-product is reformed to clean syngas energy.

The 3R plant is a technology system whereby thermal process treatment for biotech specific solid carrier manufacturing and biotechnological systems are integrated and built together into one industrial unit.

The ABC mineral solid carrier is characterized by having a grain size of around <5 mm, macroporous structure, specific surface internal area of $< 120m^2/g$ and does not contain heavy metal or organic/inorganic contamination which can inhibit the microbial activity. The main characteristics of the carrier material are variable according to different strain-specific microbiological substance need within a wide range by the special carbonization manufacturing process.

Natural soil microorganisms (bacteria and fungi) with P mobilizing and biocontrol activity have been successfully selected and specific fermentation processes have been developed. The different microorganisms, having strain and char solid carrier-specific requirements, result in different strain-specific optimizations within the process.

Innovative two-stage SSFF (solid substrate fermentation and formulation) technology and a field demonstration plant have been developed for effective formulation of the microbiological substances on a high phosphorous-based ABC mineral carrier. SSFF field demonstration units have been developed, scientifically modelled and engineering-designed, and basic frame equipment has been installed in Hungary. The developed SSFF technology and pilot plant have the advantages of superior productivity both for fungi and/or bacterial agents, advanced automation and totally closed technique, reduced energy requirement, zero-waste water output and improved product recovery. Different alternative types of food industrial by-products have been used as a C/N source for the liquid phase deep tank fermentation process. The inoculum product from liquid phase fermentation has been adapted to bone char mineral solid fermentation. The fermentation process has been optimized economically towards industrial production scenarios in the European Union and in the United States.

An economically effective nutrient and two-stage fermentation and formulation strategy has been developed for effective production of the 3R-AGROCARBON substance. The total economics has played key role in the fermentation medium development.

The different microorganisms produced different levels of colony-forming units (cfu) per gram of ABC mineral, ranging from $3x10^7$ to $5x10^8$. Improved growth during optimization of the SSFF process have been measured, as well as high survival rate after drying. A high percentage of effective and germinable propagules has been achieved with a long shelf life.

The most important experience we have achieved is the finding that there is a very significant difference of applicable results from micro-scale in vitro laboratory versus large-scale "product-like" optimization conditions, both for thermal and biotechnology processes.

It has been shown that in the realm of the natural, environmental and agricultural sciences most "product-like" scale RTD and test programmes will provide progressive and field application-oriented results, while theoretical and in vivo laboratory scale test results may lead to incorrect conclusions towards wrong development direction. Our results confirm that the large-scale laboratory fermentation and formulation conditions with industrial type nutrient and field application strategy can provide true value results that lead to successful scale-up and economical industrial application.

Figure 2. Fermentor.

The results of the SSFF tests clearly confirmed our assumption that properly produced ABC mineral is an advantageous support material for microbial colonization. Enumeration of colony-forming units within the 3R-AGROCARBON products and microscopic follow-up was found to be an easy and valuable method for quality control during and after the production of ABC with microorganisms. Improved growth during optimization of the SSFF process could be measured, as well as survival during the drying process.

The following 3R-AGROCARBON scale-up "product-like" conditions have been applied:

- Micro-scale laboratory fermentation optimization: Liquid; <5 litre; Solid: <5kg (SSFF semi sterile)
- Small-scale laboratory fermentation optimization: Liquid; <12 litre; Solid <15kg (SSFF semi sterile)
- Medium-scale laboratory fermentation optimization: Liquid; <160 litre; Solid <150kg (SSFF semi sterile)
- Large-scale lab: Liquid < 500 litre (SSFF semi sterile) "product like" test; Solid <500kg

Based on the advanced scientific RTD and product-like tests, industrial-scale ABC carbonization system and integrated SSFF have been industrially designed in detail by Edward Someus, for production of 3R-AGROCARBON products with input capacity of 30,000 m^3/year (4 m^3/h). The 3R-AGROCARBON has been successfully tested with formulations of individual fungus and bacterial strains, formulations of strain consortia and co-composting with ABC mineral/charcoal containing abundant microorganisms.

Table 14 present the inorganic analyses of the ABC mineral. Table 15 presents the inorganic element analysis of ABC mineral material with X-ray fluorescens spectrophotometer after digestion with $LiBO_2$(mg/kg) indicating change of inorganic content under the different carbonisation temperatures.

Table 14. Inorganic element analysis of ABC mineral material with X-ray fluorescents spectrophotometer after digestion with $LiBO_2$ (mg/kg)

Elements	Sample number		
	019 SV	022 SV	SS20
P	122,000	129,000	130,000
Ca	320,000	333,000	330,000
Mg	6,000	6,350	6,880
K	1,670	1,920	917
S	<200	<200	<200
Mn	5	3	2
Fe	85.7	71.4	143
Zn	173	185	170

Table 14. (Continued)

Elements	Sample number		
	019 SV	022 SV	SS20
Ba	<10	<10	<10
Pb	<2	<2	<2
Cu	3	3	4
Cr	<10	<10	<10
As	<2	<2	<2
Ni	<5	<5	<5
Cd	<2	<2	<2
Co	<10	<10	<10
Sr	138	151	136

Table 15. Inorganic element analysis of ABC mineral material with X-ray fluorescents spectrophotometer after digestion with $LiBO_2$ indicating change of inorganic content during the different carbonisation temperature (mg/kg).

Elements	Sample number (Carbonization temperature)			
	019 (500°C)	020 (550°C)	021 (600°C)	022 (700°C)
P	127,000	121,000	134,000	132,000
Ca	327,000	308,000	341,000	333,000
Mg	5,770	5,470	6,590	5,940
K	1,830	2,330	2,500	2,330
S	<200	<200	<200	<200
Mn	6	5	5	3
Fe	92.9	107	100	85.7
Zn	184	182	199	186
Sr	138	121	148	136
Ba	<10	<10	<10	<10
Pb	<2	2	<2	<2
Cu	3	2	4	3
V	<10	<10	<10	<10
Cr	<10	<10	<10	<10
Co	<10	<10	<10	<10
Ni	<5	<5	<5	<5
As	<2	<2	<2	<2
Mo	<2	<2	<2	<2
Cd	<2	<2	<2	<2
Hg	<2	<2	<2	<2

The innovative 3R-AGROCARBON method and technology is international intellectual property right "IPR" protected and a patented original solution, invented and designed by the Swedish environmental engineer Edward Someus.

Chapter 24

3R-AGROCARBON Application Targets, Concepts and Priorities

Humus is that fraction of the soil medium which aids in making minerals available to plant rootlets. The principle components of humus are a group of humic acids which have very unique properties, acting as "oscillating catalysts." With higher levels of humus in the soil, it takes fewer nutrients to produce minerals. Humus is produced in the soil by the proper decay of animal and plant matter. The principle component of humus is carbon. In order to build humus levels in soil, the proper ratio of water, air, residue, fungi, bacteria, protozoa, algae and yeasts is needed. In this context, 3R-AGROCARBON efficiently promotes the restoration and maintenance of the complex soil systems, promotes humus building and mineral mobilization towards plant availability, and naturally balances the soil life with positive effect for all of the interacting soil components. Versus this natural process, the use of acid-and salt-based fertilizers result in lower and over time totally destroy the amount of humus in the soil.

Natural soil ecological systems may lose carbon. When pastureland is converted to human agricultural cultivation land, soil carbon levels may decline. In healthy soils, carbon exists as long, sticky string-like molecules. Ploughing and burning off are two farming practices that reduce soil carbon levels. These strings twist around individual soil particles and literally bind them together. Soil microorganisms tend not to bother consuming these large, unpalatable molecules, preferring fresh or rotted plant matter – the stems, roots and other plant parts which over time become incorporated into the soil. But if the soil loses this plant content (because the stubble is burnt or removed), the soil microorganisms have no choice but to make a meal of the carbon molecules. Once the carbon is gone, the structure of the soil breaks down, making it difficult to retain water and nutrients. Carbon levels in forest soils are usually much higher than those under agriculture.

3R-AGROCARBON Carbon Economy

The two major application concepts in low input and/or organic farming are as follows:

1) Low Dose Natural Restoration By 3R-AGROCARBON: restoration of soil natural balance by enhancing the symbiosis of the soil elements by directed introduction of

selected mineral mobilization microorganisms immobilized on specific surface-modified mineral solid carriers. In this context, Low Dose of C content 3R-AGROCARBON is introduced, 400 kg/ha to 1250 kg/ha (0,1% v/v), or where it is needed a higher dose may be applied. The 3R-AGROCARBON natural restoration soils develop into organic communities that are capable of growing biotic culture towards truly living earth, renewing the soil and sequestering carbon. The combined effects are plant growth promotion with biological control by-effect against soil-borne plant pathogens, and natural fertilization of soil. In case soils are rich in P but in bound form unavailable to plants, the P mobilization selected and formulated 3R-AGROCARBON will enhance the mobilization of the in situ minerals as well for availablity for plants. The additional effect of the 3R-AGROCARBON application is the stimulation of the natural defense mechanisms of the plants.

2) High Dose Refertilization 3R-AGROCARBON fertile soil additive: refertilization of low-grade unfertile soil by 3R-AGROCARBON COMPOST (carbon-mineral adapted microbiological substance with specific compost blend) to anthropogenically (man-made) modify and rebuild the soil structure of the soil towards high C content organic matter in its top horizon approx. 20 cm depth, but without hydromorphic (water saturation) characteristics. In this context, for specific cases. High dose of C content plant-based 3R-AGROCARBON may also be introduced, 10 tons/ha to 25 tons/ha (2% v/v) or even more, subject to economy versus achieved results considerations. The combined effects are refertilization of soil, biological control against soil-borne plant pathogens, and plant growth promotion.

Chapter 25

Development of Alternative 3R AGROCARBON Strategies for Different Climatic, Environmental and Soil Conditions

The strategy of refertilization is the production of specific 3R-AGROCARBON Fertile soil additive products by controlled co-composting of agricultural and food industrial organic waste with addition of 3R-AGROCARBON microbiological substance adapted to 3R specific high-phosphorous ABC mineral and wood charcoal carriers. The introduced microorganisms support the decomposition and mineralization of the organic waste material, whereas the produced different organic compounds enhance the solubilization of the P from the apatite structure of the ABC mineral by the following mechanisms.

During the decomposition of organic waste materials (from agriculture and food industry), the 3R-AGROCARBON introduced microbial consortium can produce a large number of organic acids which enhance the solubilization of the apatite content of the ABC mineral, resulting in a specific compost material rich in available P and other key minerals and organic contents.

The enhancement of P release from apatite seems to be a function of the acidification of the P mineral by organic acids and, more importantly, their chelating ability on calcium (Ca), iron (Fe) and aluminium (Al) [78]. The greater ability of organic acids, compared with mineral acids of comparable strength, to release P from mineral, and the direct evidence of their chelating ability, have been documented [49]. Another important factor in the release of P is the participation of the OH groups in the organic acids. For example, it has been shown that citric acid with three carboxyl (COOH) groups and an OH group was able to dissolve more P from mineral than cis-aconitic acid with three carboxyl groups but without the OH group [54]. Fulvic acid is the most reactive of the humic substances in adsorbing significant amounts of Ca^{2+} and releasing H^{+} ions, thereby enhancing bone char mineral dissolution. Humic acid may form complexes with P and Ca, and create a sink for further dissolution of the mineral [86].

The application of humic substances to soil also makes more P available to plants by competing for, and by forming a protective coating over, soil phosphate-sorption sites. An additional benefit that accrues from the application of phospho-compost is the movement of dissolved P to a greater soil depth, which provides a larger soil volume for P uptake by plants

[25]. Increasing soil acidity, high cation exchange capacity, low levels of calcium (Ca) and phosphate in solution and high organic matter content favour apatite dissolution [25].

Through 3R-AGROCARBON composting, naturally-occurring decay mechanisms are controlled and encouraged to break down organic matter into compost, a relatively stable organic product that improves soil structure, aeration, and water and mineral nutrient retention.

Humus and partly decomposed soil organic matter can greatly increase the water- and nutrient-holding capacities of sandy soils [21]. For different climatic, environmental and soil conditions, alternative 3R-AGROCARBON strategy can be developed, including the enhancing of locally selected specific microbiological fungus and bacteria strain cultivations. Table 16 demonstrates the different 3R-AGROCARBON solutions and priority issues.

Table 16. 3R-AGROCARBON solutions and priority issues

3R-AGROCARBON Priorities	3R-AGROCARBON Objectives
Nutrient management and sustainable use of natural resources	Increase production with regard for the environment. Increasing nutrient use efficiency and agriculture and food industrial by-products and waste recycling through integrated nutrient sources management in cropping systems. The ABC mineral provides not only P for plant growth, but also secondary organic and inorganic nutrients, including calcium, magnesium and micro-nutrients in combination with organic compost and soil microbiological substances.
Transformation of waste into high added-value products	The 3R-AGROCARBON transformation of food chain waste into a high added-value healthy and safe product is technically, environmentally and economically very effective. 3R-AGROCARBON performs integrated carbonization and biotechnological recycling of mineral and/or carbon biotechnological carriers from organic and inorganic agricultural by-products.
Natural and environmentally friendly phosphorous supply for plants	Continuously maintaining adequate P level for plants by application of slow-release natural and environmentally friendly P fertilizers. Balanced and adequate supply of phosphorous mineral nutrient.
Carbon sequestration	An important result of the 3R-AGROCARBON technologies will turn carbons in waste streams into useful products, in many cases used as soil improvement products, which in fact is enforcing the carbon sink for the long term. The Kyoto Protocol and the US Biochar Provisions S.1884 Salazar Act highlights that soil is a major carbon store which must be protected and increased where possible.
Preventing and reversing soil degradation and increasing soil fertility	Refertilization of low grade - unfertile soil by 3R-AGROCARBON Compost (carbon-mineral adapted microbiological substance with specific compost blend). Providing integrated solution for biological control, plant

	growth promotion and natural fertilization, while supporting the restoration of soil natural balance.
Reduced fossil consumption and sustainable greenhouse gas emission management	Energy-saving integrated thermal and biological utilization and recycling organic waste and recovery of energy from it can preserve virgin materials and reduce the use of fossil fuels (so reducing greenhouse gas emissions). By increasing the amount of organic waste that is recycled, composted or has energy recovered, there is considerable scope for reducing greenhouse gas emissions from the waste produced in the agricultural, aquaculture and food industry sectors.
Minimizing or substitution of external agriculture chemicals and artificial fertilizer inputs	In order to maintain sustainable agriculture and production of safe/healthful food, the high input of chemicals and artificial fertilizers should be minimized or even substituted by natural products towards low-input and organic farming.

Chapter 26

SUSTAINABILITY OF 3R-AGROCARBON TECHNOLOGY

- ENERGY SUSTAINABLITY = uses minimum energy for agro product recycling and processing; all energy is recycled and reused.
- ECONOMICAL SUSTAINABLITY = high cost efficiency and economical value creation for farmers.
- MARKET AND CONSUMER SUSTAINABLITY = "Win-Win" for all stakeholders from producers to consumers, where the enduser is the main winner, all in order to receive very high quality and purity in nutritious crop food products on the table for lower cost.
- ENVIRONMENTAL SUSTAINABILITY = recycles and reuses waste materials to added-value products by an advanced Zero Emission carbonization process solution. Environmentally compatible agriculture practices maintain the nutrient balance of the soil.
- TECHNICAL SUSTAINABILITY = industrialized production system with controlled output quality.
- LEGAL SUSTAINABIILITY = the 3R technology operations and product applications are designed to meet the EU, USA, Canadian, Australian and Japanese norms, standards and Authority permit requirements.
- CLIMATE SUSTAINABILITY = carbon negative process

3R-AGROCARBON BENEFITS

- Overall Benefits

  - Mitigation of mineral and nutrient deficiency stress of the food crops, natural balance and functionality restoration of degraded continental agro soils in temperate climatic regions, with controlled microbiological activity and precision farming nutrient supply for sustainable, improved, economical and ecological food crop production.
  - Prevention of nutrient loss, biochar may help to reduce nutrient run-off from soils and the associated problems of eutrophication.
  - Support beneficial soil life forms towards balanced living soil
  - Soil conditioning

- The increase in available moisture observed in sandy soils make biochar a useful tool in the reversal of desertification.
- Restoration of degraded lands by increasing soil carbon
- Increasing soil carbon storage
- Improved soil moisture capture and retention

- Environmental Benefits

    - Efficient use of nutrient inputs for efficient GHG mitigation
    - Providing more conservation-oriented management practices
    - Carbon sequestration of degraded and eroded sites and areas
    - Supporting environmentally-orientated farming, development of environmentally compatible agriculture practices with maintaining the nutrient balance of the soil
    - Prevention of groundwater pollution
    - Less N_2O (very powerful GHG), even for same rate of fertilizer application-
    - Reducing direct and indirect energy use to avoid carbon emissions
    - Increasing renewable energy production to avoid carbon emissions. Biomass energy can be used to avoid greenhouse gas emission by providing equivalent energy for heat and electricity generation, and for transportation fuel. If biomass is harvested and burned, and the same area replanted or regenerated, there are no net carbon emissions over the harvest cycle.

- Economical Benefits

    - Improved farm income per farmer from increased tons/hectare, which leads to rural social improvements and gender equality
    - Reduced per hectare expenditures on fertilizers, therefore less pressure on increasingly limited oil and natural gas supplies, leading to lowered fuel costs
    - Improved rural employment opportunities
    - Combined heat and power harvested from bioenergy as a byproduct with charcoal production from biomass
    - More available farm and forest residue to support more biofuel utilization, helping on both CO_2 emission reduction and peak oil/peak gas supply constraints
    - Lowered disaster insurance costs (and similar for areas with reduced or increased rainfall, etc.)

- Application Benefits

    - Improved protection of subsurface water resources from less fertilizer runoff
    - Nutrient balance and slow release fertilizers: using nitrogen more efficiently means better matching its availability to plant needs. Remaining nitrogen is released by the charcoal slowly into the soil for the plants to absorb.
    - Biological and chemical methods for manipulating soil microbial processes to increase efficiency of nutrient uptake, suppress nitrogen emissions, and reduce leaching
    - Promotion of increased plant growth

- Less malnutrition and better health conditions from supplying more food crops per hectare
- Increased use of all the non-dispatchable renewables (wind, PV, biomass, etc.), as biochar facilities provide much of the required backup (heaviest char and electrical output during peak electrical demand periods—in the West it is July and August for air-conditioning)
- Build-up of soil organic matter that accompanies carbon sequestration provides many environmental, social, and economic co-benefits.
- The charcoal stimulates indigenous arbuscular mycorrhizal fungi in soil and thus promotes plant growth.
- Weed control and suppression by development of stronger and larger root system of the host plants
- The addition of fine charcoal in compost preparation improves the process by improving aeration, water retention probably by providing microorganisms refuge and adsorption of metabolites, and metabolite from microorganisms entrapped in the charcoal structure.
- Charcoal and its cellular structure may provide habitat for this beneficial bacterium.
- Porous substance with high water and air holding capacity, a suitable habitat for some microbes and plant growth, and a good material for soil amendment, absorption of chemicals and humidity control
- High alkalinity neutralizes acidic soil and improves chemical components of soil and selection of microorganisms
- Non organic matter exclusion of saprophytes and propagation of autotrophic and symbiotic microorganisms, free-living nitrogen-fixing bacteria, root nodule bacteria and some AM
- Charcoal as media for immobilizing useful microorganisms (spores of bacteria, root nodule bacteria, mycorrhizal fungi and nitrogen-fixing bacteria)
- Charcoal seems to assist microbial activity by having a porosity that provides a favourable microhabitat, weak alkalinity and by being a substrate unfavourable for saprophytes. AM fungi easily extend their extraradical hyphae into charcoal buried in the soil and sporulate in the particles, encouraging effects of charcoal on AM colonisation. The role of mycorrhizal fungi is particularly important in carbon cycling, because they pump the carbon compounds out of the root into a massive network of fine fungal filaments in the soil, where it becomes available to other microbes and also to larger soil organisms like worms, mites and insects. In return, the fungus gathers phosphorus from the soil and delivers it to the plant, helping the plant to grow better. Arbuscular mycorrhizal fungi, found living on plant roots around the world, appear to be the only producers of glomalin. The fungi use carbon from the plant to grow and make glomalin. In return, the fungi's hairlike filaments, called hyphae, extend the reach of plant roots. Hyphae function as pipes to funnel more water and nutrients—particularly phosphorus—to the plants. Higher levels of glomalin give greater water infiltration, more permeability to air, better root development, higher microbial activity, resistance to surface sealing (crusts) and erosion (wind/water). Benefit of glomalin: increased aggregate stability which leads to better soil structure which, in turn, leads to better plant production. As a glycoprotein, glomalin stores carbon in both its protein and carbohydrate (glucose or

sugar) subunits. Glomalin is causing a complete re-examination of what makes up soil organic matter.

- Important by-product generated in the process is fine charcoal that, in some cases, represents up to 15% of the produced charcoal-addition of fine charcoal in compost preparation improves the process by improving aeration and water retention and probably by providing microorganisms refuge and adsorption of metabolites.

Chapter 27

Organic Agrocarbon Farming

The world organic market has been growing by 20 percent a year since the early 1990s, as concern has grown about food quality and safety. Consumers consider organic products to be safer for health due to the absence of pesticide residues. North America and Europe account for 97 percent of global organic food and drink sales, but nearly half of the world's organic farmland is found in Asia, Australia and Latin America. Consumer concerns about residues of chemosynthetic pesticides and plant growth regulators continue to be a main driver for the increase in demand for foods from organic and other low input systems.

Characteristics and Definition of Organic Farming

Organic agriculture uses holistic production management systems that promote and enhance agro-ecosystem health, including biodiversity, biological cycles, and soil biological activity. Organic production systems are based on specific and precise standards of production that aim at achieving optimal agro-ecosystems thatare socially, ecologically and economically sustainable. Terms such as "biological" and "ecological" are also used in an effort to describe the organic system more clearly. Requirements for organically produced foods differ from those for other agricultural products in that production procedures are an intrinsic part of the identification and labelling of, and claim for, such products [127].

This is accomplished by using, where possible, cultural, biological and mechanical methods, as opposed to using synthetic materials, to fulfil any specific function within the system. An organic production system is designed to [127]:

- enhance biological diversity within the whole system;
- increase soil biological activity;
- maintain long-term soil fertility;
- recycle wastes of plant and animal origin in order to return nutrients to the land, thus minimizing the use of nonrenewable resources;
- rely on renewable resources in locally-organized agricultural systems;
- promote the healthy use of soil, water and air as well as minimize all forms of pollution thereto that may result from agricultural practices;
- handle agricultural products with emphasis on careful processing methods in order to maintain the organic integrity and vital qualities of the product at all stages;

- become established on any existing farm through a period of conversion, the appropriate length of which is determined by site-specific factors such as the history of the land, and type of crops and livestock to be produced [127].

Organic agriculture is claimed to be the most sustainable approach in food production. It emphasizes recycling techniques and low external input and high output strategies. It is based on enhancing soil fertility and diversity at all levels and makes soils less susceptible to erosion.

Benefits of Organic Farming

- Organic farming is considered to be environmentally beneficial partly due to the ban on synthetic fertilisers and pesticides One important reason for the increasing demand for foods from organic and low input production systems is the environmental, biodiversity and animal welfare benefits associated with these systems.
- Organic agriculture has considerable potential for *reducing emissions of greenhouse gases*. The global warming potential of organic farming systems is considerably smaller than that of conventional or integrated systems when calculated per land area.
- Organic agriculture techniques can contribute significantly to *sequestration of* CO_2 *in the soil.*
- Organic farming uses *less energy* — both per unit area and per unit of yield — and produces less extraneous waste, such as packaging materials for chemicals. Organic agriculture reduces energy requirements for production systems by 25 to 50 percent compared to conventional chemical-based agriculture.
- *There are reduced nitrous oxide emission rates* in the organic farm.
- The application of improved agricultural techniques *stops soil erosion and converts carbon losses into gains.*
- Organically managed soils have significantly higher organic matter content.
- Organic farming systems are *highly adaptive to climate change due to soil fertility-building techniques and high degree of diversity.*
- Organic farming has a positive effect on biodiversity. Organic agriculture systems are *built on a foundation of biological diversity*. Enhanced biodiversity reduces pest outbreaks. Diversified agro-ecosystems reduce the severity of plant and animal diseases, while improving utilization of soil nutrients and water.
- Organic farming *encourages soil organisms*, which can produce many compounds that help plants, including substances such as citrate and lactate that combine with soil minerals and make them more available to plant roots. The presence of these microorganisms at least partially explains the trend showing a higher mineral content of organic food crops.
- Continued use of organic fertilizers results in *increased soil organic matter, reduced erosion, better water infiltration and aeration, and higher soil biological activity* as

the materials decompose in soil and in increased yields after the year of application (residual effects).

- The organic treatments result in a *higher soil fertility capacity* and in *crops with higher quality protein, a higher starch content, and a greater ability to tolerate stressful conditions and long-term storage in comparison with the chemosynthetic mineral treatments.*
- *Organic crops contain higher levels of nutrients*, including plant secondary metabolites, substances that may be present at higher levels in organic food; organic crops contain 10–50% more secondary metabolites than conventional equivalents. This may be because fertilizers applied to conventional plants supply a surfeit of nutrients, encouraging the plant to channel more energy into growth, rather than defending against pests.
- *Organic food is free from pesticides, which are banned or severely restricted under organic regimes.*
- Productivity in sustainable agriculture, especially in organic agriculture, is enhanced by many indirect measures based on improving soil fertility and stimulating the roles of plants and microbes in natural soil processes. These can be based on symbiotic and asymbiotic nitrogen fixation and exploiting soil phosphorus and water resources by symbiotic mycorrhizal fungi.

Chapter 28

Protecting Soil Resources and Combating Climate Change

There is an urgent need to take effective measures to control GHG emissions from human activities, including agricultural production, with special attention to the carbon dioxide and methane emissions from anaerobic decomposition of livestock. The atmospheric CO_2 has increased from 280 ppm in 1750 to 367 ppm in 1999, and today's CO_2 concentrations have not been exceeded during the past 420,000 years. The industrialised countries continue to struggle with how to reduce greenhouse gas emissions by 60 percent by the year 2050, the target recommended by the Kyoto Protocol. Soil organic carbon is an important pool of carbon in the global biogeochemical cycle. The total amount of organic carbon in soils is estimated to be 2011 Gt C, which constitutes about 82 percent of the global organic carbon in terrestrial ecosystems [113].

The new U.S. legislation of July 26, 2007, act S.1884 (by Senator Salazar) on biochar provisions, promotes the commercial development of technologies that will simultaneously create clean, renewable energy from agricultural and forestry biomass products, while protecting and restoring soil resources and helping to address global climate change. Unlike most carbon-neutral biomass energy systems, aerobic biochar technology, such as 3R-AGROCARBON, is carbon-negative: it removes net carbon dioxide from the atmosphere and stores it in the form of stable soil carbon 'sinks', improving soil fertility, water retention, productivity and crop yields. More and more research shows that soils amended with the char have very beneficial effects on crop growth. The enhanced nutrient retention capacity of biochar-amended soil not only reduces the total fertilizer requirements but also the climate and environmental impact of croplands. Char-amended soils have shown reductions in nitrous oxide emissions and reduced runoff of P into surface waters and leaching of nitrogen into groundwater. As a soil amendment, biochar significantly increases the efficiency of and reduces the need for traditional chemical fertilizers, while greatly enhancing crop yields. The method for sequestering carbon in the soil involves a gentler way of farming the land, and the potential for sequestering carbon is huge. The 3R-AGROCARBON is reducing impacts on the climate by reducing its own carbon footprint in the form of carbon offsetting.

Chapter 29

Product Permit EU Authority and Commercial Field Cultivation Test Results for Validation and Demonstration of 3R-AGROCARBON Agronomic and Nutrition Effectiveness

During the course of the applied scientific RTD and application oriented optimization, the 3R-AGROCARBON products have been successfully multiple tested for several years under different climatic, soil and agro-industrial production conditions in Hungary, Israel, Italy, Germany and the Netherlands. This validation test case study below is executed by the independent and accredited Agricultural Office, Plant Protection and Soil Conservation Directorate, the authority in Hungary, as part of the EU permit procedures for plant growth promoter with biological control by-effect [2]. Both our own Europe-wide past years' tests results and the authority tests clearly indicate positive 3R-AGROCARBON product efficiency under different field conditions, on which true tests have been executed as of agricultural industrial practice. It is estimated that by the end of 2008, the authority permit tests will be completed and product permit granted.

Authority 2007 Field Test Conditions

Test plant:

Test Program I.

- Tomato: *Lycopersicon lycopersicum*
- Sweet pepper: *Capsicum annuum*

Test Program II.

- Broccoli: *Brassica oleracea Italica*

Test soil type:

The test location and soil type was the same during the Test Programs I and II

Before selecting the experimental plots, soil samples were taken for determination of soil nutrient content. The results indicated reach macro- and micro element content of the test soil (see Table 17.)

Table 17. Measurement of nutrient condition of the test soil (soil type: chernozem with lime deposit) (3R-AGROCARBON Authority Test Programs I-II)

Measured soil parameter	Values
pH_{KCl}	7.49
Salt content (%)	<0.02
$CaCO_3$ (%)	9
Humus content (mass %)	3.50
$NO_2 + NO_3$-N (mg/kg)	18.4
P_2O_5 (mg/kg)	195
K_2O (mg/kg)	228
Na (mg/kg)	98.6
Mg (mg/kg)	181
Cu (mg/kg)	2.88
Zn (mg/kg)	1.86
Mn (mg/kg)	44.4
SO_4-S (mg/kg)	11.9

Arrangement of test plants and repetitions in Test Programs I and II: Random block arrangement was applied with 4 repetitions.

Application method of 3R-AGROCARBON products and doses: Before planting 3R-AGROCARBON product and artificial fertilizers were dispersed and worked it immediately with rotational hoe.

The following doses were applied during Test Programs I and II. (see Tables 18 and 19).

Test Program I.

ABC mineral without microbial adaptation: 600kg/hectare
Microbial (3R-T10) adapted ABC mineral: 400kg/ha, 600kg/ha and 1000kg/hectare

Test Program II.

ABC mineral without microbial adaptation: 600kg/hectare
Microbial (3R-T10) adapted ABC mineral: 400kg/ha, 600kg/ha and 1000kg/hectare
Microbial (3R-T10) adapted ABC mineral (3 month stored): 600 kg/hectare

Table 18. Data of the different 3R-AGROCARBON treatments (3R-AGROCARBON Authority Test Program I)

N°	Treatment	Dose kg/ha	Time	Fenophase	Mode of treatment
1	Untreated Control	-	-	-	-
2	ABC mineral	600	2007.06.13	Before planting	Manual spread to soil
3	ABC mineral+ Trichoderma sp.	400	2007.06.13	Before planting	Manual spread to soil
4	ABC mineral+ Trichoderma sp.	600	2007.06.13	Before planting	Manual spread to soil
5	ABC mineral+ Trichoderma sp.	1000	2007.06.13	Before planting	Manual spread to soil

Table 19. Data of the different 3R-AGROCARBON treatments (3R-AGROCARBON Authority Test Program I)

N°	Treatment	Dose kg/ha	Time	Fenophase	Mode of treatment
1	Untreated Control	-	-	-	-
	ABC mineral	600	2007. 08. 07.	Before planting	Manual spread to soil
	ABC mineral +Trichoderma sp. (3 month stored)	600	2007. 08. 07.	Before planting	Manual spread to soil
	ABC mineral +Trichoderma sp.	400	2007. 08. 07.	Before planting	Manual spread to soil
	ABC mineral +Trichoderma sp.	600	2007. 08. 07.	Before planting	Manual spread to soil
	ABC mineral +Trichoderma sp.	100	2007. 08. 07.	Before planting	Manual spread to soil

Fertilization Practice

Only mineral fertilizations were applied. No adjunctive organic matter based fertilizers were adopted. To every parcel, with the exception of the control treatment, 20 kgs of N active ingredient in the form of ammonium-nitrate and 200 kgs of K_2O ingredient in the form of potassium sulphate nutrients per ton were added.

Irrigation Condtition

Because of the extraordinarily warm and drought-like weather during Test Programs I and II, regular season artifical irrigation control was required for providing sufficient water for the plants.

Assessment of the 3R-AGROCARBON Qualitative and Quantitative Effects

During Test Programs I and II. regular physiological and phenological observations were performed. The tomatoes were cultivated until the first harvest period; sweet peepers until the third harvest periods while broccoli until the total bearing; and investigated for the following important parameters: measurement of total weight of fruits, soil and plant leaves nutrient content change after the 3R-AGROCARBON treatments.

Results

Sweet pepper

At the first harvesting, the size and yields of the three types of 3R-AGROCARBON treated plants were three to five percent higher compared to the untreated control, but differences between the treatments were not significant (see Table 20).

Table 20. Measurement of fruit weight in sweet pepper at first harvest (3R-AGROCARBON Authority Test Program I)

N°	Treatment	Dose (kg/ha)	Repetitions I.	II.	III.	IV.	Average kg/parcel	Control %
1	Untreated control	-	3.308	3.487	3.985	3.258	3.510	100
2	ABC	600	3.747	3.033	3.987	3.700	3.617	103.1
3	ABC+ Trichoderma sp.	400	3.300	3.790	3.986	3.517	3.648	104.0
4	ABC+ Trichoderma sp.	600	4.711	3.433	3.114	3.496	3.689	105.1
5	ABC+ Trichoderma sp.	1000	3.114	3.192	3.789	4.591	3.672	104.6

At the second harvesting the tendency was the same as after the first period, but the differences were not significant because of the high variance. However, it can be seen that the largest quantity of fruits were measured in the 1000 kg/ha dose of 3R-AGROCARBON treatment. (See Table 21.)

The results of the last measurement (see Table 22) were remarkable from different viewpoints. It can be clearly seen that the positive tendency was the same as received after the first and second harvesting period. The effect of the blank ABC and low dose 3R-AGROCARBON treatment were similar, while in the 600 and 1000 kg 3R-AGROCARBON treatments the yields significantly exceed the untreated control.

Table 21. Measurement of fruit weight in sweet pepper at second harvest (3R-AGROCARBON authority Test Program I)

N°	Treatment	Dose (kg/ha)	Repetitions I.	II.	III.	IV.	Average kg/parcel	Control %
1	Untreated control	-	6.224	5.791	8.991	8.670	7.419	100
2	ABC	600	8.135	7.258	7.525	7.958	7.719	104
3	ABC+ Trichoderma sp.	400	7.606	8.777	6.809	7.795	7.747	104.4
4	ABC+ Trichoderma sp.	600	7.370	8.005	7.736	8.180	7.823	105.4
5	ABC+ Trichoderma sp.	1000	8.269	7.716	7.724	8.523	8.058	108.6

Table 22. Measurement of fruit weight of sweet pepper at third harvest (3R-AGROCARBON authority Test Program I.)

N°	Treatment	Dose (kg/ha)	Repetitions I.	II.	III.	IV.	Average kg/parcel	Control %
1	Untreated control	-	4.020	3.330	3.540	3.580	3.618	100
2	ABC	600	3.120	3.820	4.040	3.920	3.725	103.0
3	ABC+ Trichoderma sp.	400	3.420	4.120	3.970	3.540	3.763	104.0
4	ABC+ Trichoderma sp.	600	3.740	4.380	3.980	3.680	3.945	109.1
5	ABC+ Trichoderma sp.	1000	4.000	4.800	4.650	4.300	4.438	122.7

These results indicated the effective and clear phosphorous and nutrient mobilization and plant growth promotion effects of the 3R-AGROCARBON treatments. The results also indicated that the effective mobilization required a minimum of three months.

The application of the specific ABC mineral increased the yields by one tonne/ha. The formulated ABC mineral carrier by *Trichoderma harzianum* sp. 3R-T10 further increased the yields in the field test with two additional tonnes/ha. The largest yields resulted from the 1000 kg/ha 3R-AGROCARBON treatment dose, where production yield increased by about 3 tonne/ha, which was a significant result. (See Table 23 and Table 24.)

Table 23. Final results of sweet pepper test yield values (3R-AGROCARBON Authority Test Program I)

N°	Treatment	Dose (kg/ha)	Repetitions				Average kg/parcel	Control %
			I.	II.	III.	IV.		
1	Untreated control	-	13.55	12.61	16.52	15.51	14.546	100
2	ABC	600	15.00	14.11	15.55	15.58	15.061	103.5
3	ABC+ Trichoderma sp.	400	14.33	16.69	14.77	14.85	15.158	104.2
4	ABC+ Trichoderma sp.	600	15.82	15.82	14.83	15.36	15.456	106.3
5	ABC+ Trichoderma sp.	1000	15.38	15.71	16.16	17.41	16.167	111.1

Table 24. Sweet pepper yield values at different treatment per 1 hectare (3R-AGROCARBON Authority Test Program I.)

N°	Treatment	Dose (kg/ha)	Repetitions				Average t/hectare	Control %
			I.	II.	III.	IV.		
1	Untreated control	-	25.80	14.01	31.45	29.53	27.696	100
2	ABC	600	28.56	26.87	29.61	29.66	28.676	103.5
3	ABC+ Trichoderma sp.	400	27.28	31.77	28.11	28.28	28.860	104.2
4	ABC+ Trichoderma sp.	600	30.12	30.12	28.24	29.24	29.429	106.3
5	ABC+ Trichoderma sp.	1000	29.29	29.91	30.77	33.16	30.782	111.1

The results of the soil nutrient analysis clearly indicated that (see Table 25) the plant available phosphorous content significantly increased. The results also indicated the effective phosphorous mobilization activity of the selected *Trichoderma* sp.

Table 25. Variation of the soil available P_2O_5 content after the different treatments in sweet pepper test (3R-AGROCARBON Authority Test Program I)

N°	Treatment	Dose (kg/ha)	Repetitions I.	II.	III.	IV.	Average mg/kg	Control %
1	Untreated control	-	190	178	180	179	181.750	100
2	ABC	600	160	180	196	209	186.250	102.5
3	ABC+ Trichoderma sp.	400	199	188	194	195	194.000	106.7
4	ABC+ Trichoderma sp.	600	209	190	184	208	197.750	108.8
5	ABC+ Trichoderma sp.	1000	251	223	193	197	216.000	118.8

The increasement of plant available Nitrogen and Potassium content is resulted from the addition of mineral fertilizers. Significantly higher (10 percent) available soil Potassium concentration (see Table 26.) has been measured in all treatetment compared to the non treated control.

Table 26. Variation of soil available K_2O content after the different treatments in sweet pepper (3R-AGROCARBON Authority Test Program I)

N°	Treatment	Dose (kg/ha)	Repetitions / mg/kg I.	II.	III.	IV.	Average mg/kg	Control %
1	Untreated control	-	223	173	263	232	22.75	100.0
2	ABC	600	261	339	226	360	296.50	133.1
3	ABC+ Trichoderma sp.	400	300	323	263	376	315.50	141.6
4	ABC+ Trichoderma sp.	600	336	284	349	292	315.25	141.5
5	ABC+ Trichoderma sp.	1000	342	270	404	279	323.75	145.3

Table 27. Variation of soil available NO_3 and NO_2 content after the different treatments in sweet pepper test (3R-AGROCARBON Authority Test Program I.)

N°	Treatment	Dose (kg/ha)	Repetitions I.	II.	III.	IV.	Average mg/kg	Control %
1	Untreated control	-	2.27	2.41	4.43	4.44	3.377	100.0
2	ABC	600	3.72	4.94	3.14	4.67	4.118	121.5
3	ABC+ Trichoderma sp.	400	4.07	4.21	3.95	3.79	4.005	118.2
4	ABC+ Trichoderma sp.	600	4.9	4.08	3.28	4.31	4.143	122.3
5	ABC+ Trichoderma sp.	1000	4.16	3.94	3.55	4.87	4.130	121.9

The results of the plant analysis indicated that 3R-AGROCARBON treatments increased the generally stable P content in the plant leaves (See Table 28). The most impressive results have been archived by 1000 kg/ha 3R-AGROCARBON treatment.

Table 28. Variance of the phosphorous content in sweet pepper leaves after different treatments (3R-AGROCARBON Authority Test Program I)

N°	Treatment	Dose (kg/ha)	Repetitions I.	II.	III.	IV.	Average m/m %	Control %
1	Untreated control	-	0.244	0.230	0.261	0.254	0.247	100
2	ABC	600	0.251	0.249	0.247	0.254	0.250	101.2
3	ABC+ Trichoderma sp.	400	0.261	0.248	0.235	0.263	0.252	101.8
4	ABC+ Trichoderma sp.	600	0.276	0.244	0.260	0.265	0.261	105.7
5	ABC+ Trichoderma sp.	1000	0.275	0.269	0.254	0.254	0.263	106.4

No significantly higher nitrogen and potassium concentration (see Table 29 and Table 30) has been measured in all treated sweet pepper leaves compared to the untreated control.

Table 29. Variation of Nitrogen content in sweet pepper leaves after different treatments (3R-AGROCARBON Authority Test Program I.)

N°	Treatment	Dose (kg/ha)	Repetitions I.	II.	III.	IV.	Average m/m %	Control %
1	Untreated control	-	3.19	2.96	3.45	2.89	3.123	100.0
2	ABC	600	3.09	2.86	3.41	3.19	3.138	100.5
3	ABC+ Trichoderma sp.	400	3.58	3.56	2.61	3.55	3.325	106.5
4	ABC+ Trichoderma sp.	600	3.58	3.41	3.01	3.29	3.323	106.4
5	ABC+ Trichoderma sp.	1000	2.97	3.41	3.49	3.63	3.375	108.1

Table 30. Variation of Potassium content in sweet pepper leaves after different treatments (3R-AGROCARBON Authority test program I.)

N°	Treatment	Dose (kg/ha)	Repetitions I.	II.	III.	IV.	Average m/m %	Control %
1	Untreated control	-	2.460	1.930	2.650	1.977	2.254	100.0
2	ABC	600	2.620	2.630	2.787	2.560	2.649	117.5
3	ABC+ Trichoderma sp.	400	3.100	2.541	2.400	2.810	2.713	120.3
4	ABC+ Trichoderma sp.	600	2.785	3.150	2.450	2.500	2.721	120.7
5	ABC+ Trichoderma sp.	1000	2.660	3.110	2.660	2.687	2.779	123.3

Tomato

The tomato test results indicated a similar effect as that of the sweet pepper field test. The tomato yield given the largest dose (1000 kg/ha) 3R-AGROCARBON treatment was 7.5 percent higher as compared to a non treated control, is equivalent to 5.2 tonnes per hectare surplus. But the lower dose (600 kg/ha) of 3R-AGROCARBON application still resulted in a

2.7 tonnes per hectare surplus as compared to an untreated control. (See Table 31 and Table 32.)

Table 31. Total tomato yield values after different 3R-AGROCARBON treatments (3R-AGROCARBON Authority Test Program I)

N°	Treatment	Dose (kg/ha)	Repetitions				Average kg/parcel	Control %
			I.	II.	III.	IV.		
1	Untreated control	-	12.64	11.60	11.20	13.10	12.135	100
2	ABC	600	10.80	12.86	13.10	12.34	12.275	101.2
3	ABC+ Trichoderma sp.	400	13.10	11.80	11.98	12.90	12.445	102.6
4	ABC+ Trichoderma sp.	600	12.80	12.58	12.20	12.88	12.615	104
5	ABC+ Trichoderma sp.	1000	13.0	13.50	11.90	13.79	13.048	107.5

Table 32. Results of tomato yields per hectare (3R-AGROCARBON Authority Test Program I)

N°	Treatment	Dose (kg/ha)	Repetitions				Average kg/ha %	Control %
			I.	II.	III.	IV.		
1	Untreated control	-	72.0	66.1	63.8	74.7	69.170	100
2	ABC	600	61.6	73.3	74.7	70.3	69.968	101.2
3	ABC+ Trichoderma sp.	400	74.7	67.3	68.3	73.5	70.937	102.6
4	ABC+ Trichoderma sp.	600	73.0	71.7	69.5	73.4	71.906	104.0
5	ABC+ Trichoderma sp.	1000	74.1	77.0	67.8	78.6	74.371	107.5

The results of the soil nutrient analysis clearly indicated that (see Table 33) the plant available phosphorous content significantly increased in the tomato test.

Table 33. Variation of soil available P_2O_5 content after the different treatments in tomato test (3R-AGROCARBON Authority Test Program I)

No	Treatment	Dose (kg/ha)	Repetitions I.	II.	III.	IV.	Average mg/kg	Control %
1	Untreated control	-	202	189	195	191	194.250	100
2	ABC	600	253	202	180	190	206.250	106.2
3	ABC+ Trichoderma sp.	400	253	211	192	207	225.000	115.8
4	ABC+ Trichoderma sp.	600	240	238	215	207	225.000	115.8
5	ABC+ Trichoderma sp.	1000	313	208	210	209	235.000	121.0

Table 34 demonstrates the increasing P content in the tomato leaves after different treatments.

Table 34 Variation of phosphorous content in tomato leaves after different treatments (3R-AGROCARBON Authority Test Program I)

N°	Treatment	Dose (kg/ha)	Repetitions I.	II.	III.	IV.	Average m/m %	Control %
1	Untreated control	-	0.279	0.331	0.280	0.288	0.295	100
2	ABC	600	0.312	0.282	0.301	0.285	0.295	100.2
3	ABC+ Trichoderma sp.	400	0.321	0.274	0.287	0.307	0.297	100.9
4	ABC+ Trichoderma sp.	600	0.325	0.295	0.318	0.277	0.304	103.1
5	ABC+ Trichoderma sp.	1000	0.289	0.314	0.327	0.292	0.306	103.7

Broccoli

Yield measurement has been executed in each experimental plot after complete maturation.

No any fitotoxic effect, development anomaly or disease development were observed during the test.

Application of ABC mineral alone did not cause significant yield increase. The largest yield resulted from the 600 kg/ha dose 3R-AGROCARBON treatment, where the 12% increase was significant result. The higher doses (1000 kg/ha) did not cause further yield increase (see Table 35.). The stored 3R-AGROCARBON product produced a smaller effect than that of the fresh product.

Table 35. Measurement of broccoli yield after the test (3R-AGROCARBON Authority Test Program II)

No	Treatment	Dose (kg/ha)	Repetitions				Average kg/parcel	Control %
			I.	II.	III.	IV.		
1	Untreated control	-	24.74	22.90	23.81	21.29	23.338	100.0
2	ABC	600	24.76	23.24	23.62	23.20	23.705	101.6
3	ABC+ Trichoderma sp. (stored)	600	21.90	23.62	28.65	24.76	24.733	106.0
3	ABC+ Trichoderma sp.	400	24.69	22.97	24.80	27.96	25.105	107.6
4	ABC+ Trichoderma sp.	600	26.86	24.95	27.35	25.30	26.114	111.9
5	ABC+ Trichoderma sp.	1000	24.95	29.03	24.30	21.98	25.067	107.4

The results of the soil nutrient analysis (see Tables, 36-37-38) clearly demonstrated the positive effect of all treatments. The increase in plant available nitrogen and potassium content resulted from the addition of mineral fertilizers (see table 37 and 38). The increasing available phosphorous content (see Table 36) evidently resulted from the P content of the ABC mineral caused by mobilisation activity of *Trichoderma* fungi. The plant available phosphorous content significantly increased in the broccoli test. The higher amount resulted from higher application doses.

Table 36. Variation of soil available P_2O_5 content after the different treatments in broccoli test (3R-AGROCARBON Authority Test Program II)

N°	Treatment	Dose (kg/ha)	Repetitions / P_2O_5 mg/kg				Average P_2O_5 mg/kg	Control %
			I.	II.	III.	IV.		
1	Untreated control	-	251.0	262.0	254.0	248.0	253.750	100.0
2	ABC	600	249.0	242.0	285.0	311.0	271.750	107.1
3	ABC+ Trichoderma sp. (stored)	600	241.0	280.0	300.0	269.0	272.500	107.4
4	ABC+ Trichoderma sp.	400	260.0	241.0	251.0	254.0	251.500	99.1
5	ABC+ Trichoderma sp.	600	252.0	261.0	270.0	261.0	261.000	102.9
6	ABC+ Trichoderma sp.	1000	310.0	252.0	307.0	256.0	281.250	110.8

Table 37. Variation of soil available NO_3 and NO_2 content after the different treatments in broccoli test (3R-AGROCARBON Authority Test Program II.)

N°	Treatment	Dose (kg/ha)	Repetitions				Average mg/kg	Control %
			I.	II.	III.	IV.		
1	Untreated control	-	5.16	4.02	5.63	5.12	4.983	100.0
2	ABC	600	6.10	5.25	7.44	7.95	6.685	134.2
3	ABC+ Trichoderma sp. (stored)	600	5.26	6.14	8.74	7.20	6.835	137.2
4	ABC+ Trichoderma sp.	400	7.13	5.30	7.14	5.79	6.340	127.2
5	ABC+ Trichoderma sp.	600	5.32	7.27	7.21	5.25	6.263	125.7
6	ABC+ Trichoderma sp.	1000	6.20	5.41	6.23	5.54	5.845	117.3

Table 38. Variation of soil available K_2O content after the different treatments in broccoli test (3R-AGROCARBON Authority Test Program II.)

N°	Treatment	Dose (kg/ha)	Repetitions / mg/kg				Average mg/kg	Control %
			I.	II.	III.	IV.		
1	Untreated control	-	219.0	273.0	220.0	226.0	234.500	100.0
2	ABC	600	250.0	242.0	234.0	274.0	250.000	106.6
3	ABC+ Trichoderma sp. (stored)	600	216.0	232.0	316.0	278.0	260.500	111.1
4	ABC+ Trichoderma sp.	400	244.0	235.0	237.0	268.0	246.000	104.9
5	ABC+ Trichoderma sp.	600	232.0	223.0	267.0	277.0	249.750	106.5
6	ABC+ Trichoderma sp.	1000	235.0	246.0	251.0	296.0	257.000	109.6

No significantly lower nitrogen concentration (see Table 39.) has been measured in all treated broccoli leaves compared to the untreated control. The phosphorous (see Table 40) and potassium (see Table 41) concentration in the broccoli leaves have been significantly increased by application of both 600kg/ha and 1000 kg/ha 3R-AGROCARBON doses.

Table 39. Variation of nitrogen content in broccoli leaves after different treatments (3R-AGROCARBON Authority Test Program II.)

N°	Treatment	Dose (kg/ha)	Repetitions				Average m/m %	Control %
			I.	II.	III.	IV.		
1	Untreated control	-	4.36	3.78	4.46	4.39	4.248	100.0
2	ABC	600	2.78	4.19	4.05	3.57	3.648	85.9
3	ABC+ Trichoderma sp. (stored)	600	4.46	3.87	3.74	4.71	4.195	98.8
4	ABC+ Trichoderma sp.	400	3.28	4.32	3.75	5.29	4.160	97.9
5	ABC+ Trichoderma sp.	600	4.82	4.14	4.40	3.35	4.178	98.4
6	ABC+ Trichoderma sp.	1000	4.67	3.35	4.09	3.95	4.015	94.5

Table 40 demonstrates the increasing P content in the broccoli leaves after different treatments.

Table 40. Variation of Phosphorous content in broccoli leaves after different treatments (3R-AGROCARBON Authority Test Program II)

N°	Treatment	Dose (kg/ha)	Repetitions				Average m/m %	Control %
			I.	II.	III.	IV.		
1	Untreated control	-	0.229	0.335	0.306	0.306	0.294	100.0
2	ABC	600	0.317	0.316	0.355	0.347	0.334	113.5
3	ABC+ Trichoderma sp. (stored)	600	0.359	0.323	0.309	0.390	0.345	117.4
4	ABC+ Trichoderma sp.	400	0.268	0.296	0.306	0.426	0.324	110.2
5	ABC+ Trichoderma sp.	600	0.371	0.359	0.331	0.346	0.352	119.6
6	ABC+ Trichoderma sp.	1000	0.400	0.296	0.372	0.371	0.360	122.4

Table 41. Variation of Potassium content in broccoli leaves after different treatments (3R-AGROCARBON Authority Test Program II)

N°	Treatment	Dose (kg/ha)	Repetitions				Average m/m %	Control %
			I.	II.	III.	IV.		
1	Untreated control	-	0.898	1.020	0.797	0.983	0.925	100.0
2	ABC	600	0.813	0.997	1.340	1.090	1.060	114.7
3	ABC+ Trichoderma sp. (stored)	600	1.140	1.180	0.742	1.260	1.081	116.9
4	ABC+ Trichoderma sp.	400	0.809	1.200	0.952	0.990	0.988	106.8
5	ABC+ Trichoderma sp.	600	1.080	1.260	0.809	1.710	1.215	131.4
6	ABC+ Trichoderma sp.	1000	1.130	1.130	1.280	1.280	1.205	130.3

Chapter 30

CONCLUSION

The test program results clearly indicated the biological effectiveness of the microbial (3R-T10) adapted ABC mineral product. The application of 3R-AGROCARBON substances result in balanced plant available mineral content of soil and balanced nutrient uptake for plants. Significant yield increase and fruit quality improvement has also been demonstrated in tomato, sweet pepper and broccoli.

Summarizing the results of the tomato and sweet pepper trials, it is clearly indicated that the developed 3R-AGROCARBON product is an effective plant natural nutrient, plant growth promoter and plant health improver with biological control effects against soil borne plant diseases. The biological efficiency of the ABC mineral carrier colonized by *Trichoderma* sp. is evident. Other fungus and bacteria strains have also been successfully tested. After the 3R-AGROCARBON treatment the soil plant available phosphorous content increased, which resulted in 11 percent higher plant yield compared to the control in average. Both untreated tomato and sweet pepper controls produced a good average yield, which indicates that under more disadvantageous soil and climate conditions, the different 3R-AGROCARBON treatments would have had even higher surplus effects on the yield. The tendency and positive effect of the 3R-AGROCARBON treatments to the soil available Phosphorous content and plant yields were similar to those observed during the sweet pepper test. After 600-1000 kg/ha 3R-AGROCARBON treatments, the plant available phosphorous content of the soil versus control was significantly increased, in average 20%.

Further authority tests with other horticultural test plants on sandy loam soil with low available nutrient content have been successfully completed in late 2008, and final EU permit dossier submitted early 2009. The results on low-quality soils so far are more than favourable and indicate a similar positive effect tendency with high efficiency than tests under different climatic and soil conditions. After the 600 kg/ha 3R-AGROCARBON treatment the soil plant available phosphorous content remarkably increased, resulting in 25 percent higher plant yield compared to the control on average, while the quality of the fruit was improved as well.

Chapter 31

FUTURE TRENDS

By identification of future trends and driving forces, which will be the framework for the agricultural economy on the horizon in 2020, economic development, agricultural and environmental policy have been considered as major impacting factors [82]. There are food supply uncertainties we face in the twenty-first century for which there are several reasons:

a) increasing use of crops for bio-energy, resulting in increased agricultural product prices
b) increasing demand for meat and milk products in the developing world
c) poor harvests around the world following droughts and floods
d) consumer awareness for higher food quality is rising, which is also supported by easy information access via the Internet, resulting in the following:
e) explosive demand for low input farming food products, whereas less and milder chemicals used, and heavy metals removed from the cultivation inputs
f) explosive demand for organic farming food products, whereas all chemicals and heavy metals removed from the cultivation inputs, same as already specified for organic farming
g) demand for safer, more nutritious and more tasty food products, for affordable and reasonably low prices. The agricultural environmental and ecological issues are further expanded and these are considered as more than ever critically important for sustainable development. These include the following:
h) comprehensive measures taken to minimize the agricultural environmental and ecological impacts towards protection and preservation of biological diversity
i) healthy soil is realized as one of the most important natural resources
j) The biochar carbon-negative [101] technology (i.e., 3R-AGROCARBON) for protecting and restoring soil resources might be an important element in addressing global climate change and at the same time increase high quality food crop production in a sustainable way in the low input and organic cultivation sectors.

The predicted future trend is that both soil and water of good quality and sufficient quantity are becoming rare resources [82]. Advanced, but economical, technological solutions are needed to combat these negative trends, which include a reduction in the levels of biodiversity. The existing social demand for aesthetically pleasing, “traditional” landscapes is further increased. Another trend is the consumer requirement that choice of food applies not only to the variety but also to the origin and mode of production of agricultural commodities. The evolution of trade and consequently of markets determines whether these commodities are sourced locally or by suppliers outside of borders. The combination of agricultural technology and markets frame the production possibilities, and evidently directly determine how the land is used, and – by extension – if it will come out of production. Finally, both employment opportunities and life-style choices are jointly having a profound effect on the migratory movements that are occurring, with the added dimension that communication and transport technology gives additional scope for the individual freedom to choose where to live.

In all of these trends, the 3R-AGROCARBON innovative technology offers sustainable and economical solutions, where the natural preservation concept from the ancient *Lithica,* 400 BC is considered: *"For all the pests that out of earth arise the earth itself the antidote supplies".*

SOURCES OF INFORMATION AND ADVICE

3R Environmental Technologies Ltd and Terra Humana Ltd are continuously developing their Web Sites, http://www.3ragrocarbon.com *and* http://www.terrenum.net *to provide information and consultation for the public. To contact Edward Someus by e-mail, write to:* edward.someus@gmail.com or edward@terrenum.net.

REFERENCES

[1] Abelson, P. H. *A Potential Phosphate Crisis.* Science, 26 March 1999: Vol. 283. no. 5410, p. 2015).

[2] Agricultural Office, Plant Protection and Soil Conservation Directorate, Directorate Fejer County, Hungary. Report on evaluation of 3R-AGROCARBON plant growth promotion material, October 15, 2007.

[3] Almaraz, R., van den Berg, E., Reich P. (1997). An Assessment of the Soil Resources of Africa in Relation to Productivity. *Geoderma,* 77, 1-18.

[4] American Chemical Society (2008). Ancient Method, 'Black Gold Agriculture' May Revolutionize Farming, Curb Global Warming. *Science Daily.* Retrieved May 9, 2008. Available from URL: http://www.sciencedaily.com /releases/2008/04/ 080410153658. htm

[5] Anwar, E. Z., Wissa, M.A. Phosphogypsum Disposal and The Environment. Available from URL: http://www.fipr.state.fl.us/pondwatercd /phosphogypsum_ disposal.htm

[6] Ball, A.S., Pretty,J.N. Agricultural influences on carbon emissions and sequestration. Powell et al. (ed), *UK Organic Research 2002: Proceedings of the COR Conference,* 26-28th, March 2002, Aberystwyth, pp. 247-249.

[7] Barber, S.A. (1995). *Soil nutrient bioavailability. A mechanistic approach.* New York, USA, John Wiley and Sons.

[8] Bellamy, P. H., Loveland, P. J., Brabley, R. I., Lark, R. M., Kirk, G. J. D. Carbon losses from all soils across England and Wales 1978-2003. *Nature,* 2005. vol. 437 pp. 245-248.

[9] Bieleski, R.L. (1973). Phosphate pools, phosphate transport, and phosphate availability. *Annu Rev Plant Physiol ,* Volume 24: 225–252.

[10] Brandt-Williams, S., Pillet, G. Fertilizer co-producrs as agricultural emternalities. Quantifying Environmental Services Used in Production of Food, Biennial Emergy Analysis Conference, publ. 2003.

[11] Brenner, J., Paustian, K., Bluhm, G., Cipra, J., Easter, M., Elliott, T., Kautza, T., Killian, K., Schuler, J., and Williams, S. Quantifying the change in greenhouse gas emissions due to natural resource conservation practice application in Iowa. Final Report to the Iowa Conservation Partnership. March, 2001, USDA Natural Resources Conservation Service.

[12] Bumb, B.L., Baanante, C.A. (1996) The Role of Fertilizer in Sustaining Food Security and Protecting the Environment. Food, Agriculture and the Environment Discussion Paper, 17. International Food Policy Research Institute, Washington, DC.

[13] Bundesministerium für Ernährung, Landwirtschaft und Verbraucherschutz. 2006. Die EU-Agrarreform–Umsetzung in Deutschland 2006. Available from URL: www.bmelv.de

[14] Chameides, W. L., Kasibhatla, P. S., Yienger, J. and Levy, H. (1994). Growth of continental-scale metro-agroplexes, regional ozone pollution, and world food production. *Science,* 264, 74–77.

[15] Comis, D. Glomalin hiding place for a third of the world's stored soil carbon. Agricultural Research. Sept 2002. *FindArticles.com.,* 09. May 2008. Available from URL: http://findarticles.com/p/articles /mi_m3741/is_9_50/ai_92589768

[16] Connett, M. (2003). The Phosphate Fertilizer Industry: An Environmental Overview, Fluoride Action Network, Available from URL: http://www.fluoridealert.org/phosphate/overview.htm

[17] Cooperband, L. (2002). Building Soil Organic Matter with Organic Amendments, *published by the Center for Integrated Agricultural Systems (CIAS)*, College of Agricultural and Life Sciences, University of Wisconsin-Madison.

[18] Council for Agricultural Science and Technology (2004). *Climate Change and Greenhouse Gas Mitigation:Challenges and Opportunities for Agriculture*. Ames, IA: CAST, 120 p.

[19] Delmas, R., Serca, D. and Jambert, C. (1197). Global inventory of NOx sources. *Nutr. Cycl. Agroecosyst.*, 48, 51–60.

[20] Doerr, S. H., R. A. Shakesby, and R. P. D. Walsh. (2000). Soil water repellency: its causes, characteristics and hydro-geomorphological significance. *Earth-Science Reviews* 51, no. 1-4 (August): 33-65.

[21] Dunn, R.A. (1994). *Soil Organic Matter, Green Manures and Cover Crops for Nematode Management.* SS-ENY-918, Entomology and Nematology Department, Florida Cooperative Extension Service, Institute of Food and Agricultural Sciences, University of Florida.

[22] Eckert, D.,Efficient Fertilizer Use Mannual - Nitrogen, Available from URL: http://64.51.188.251/Assets/pdffiles/efu/Nitrogen.pdf

[23] EcosSanres Fact Sheet 4, Closing the Loop on Phosphorus. Available from URL: http://www.ecosanres.org/pdf_files/Fact_sheets/ESR4lowres.pdf

[24] FAO. (1984). *Fertilizer and plant nutrition guide.* FAO Fertilizer and Plant Nutrition Bulletin No. 9. Fertilizer and Plant Nutrituion Service, Land and Water Development Division.

[25] FAO. (2004). *Use of phosphate rocks for sustainable agriculture.* FAO Fertilizer and plant nutrition bulletin 13. Technical editor: E. Zapata.

[26] Firbank, L. Sowing the seeds of farming's future. BBC NEWS, Science, November 13, 2007.

[27] Francl, T., Nadler, R., and Bast, J. (1998). Impact of the Kyoto Protocol on Agriculture. American Council for Capital Formation, policy conference presentation sponsored by the ACCF Center for Policy Research.

[28] Funderburg, E., Why are nitrogen prices so high? Available from URL: http://www.noble.org/Ag/Soils/NitrogenPrices/Index.htm

[29] Gellings, C.W., Parmenter, K.E.. Energy Efficiency in Fertilizer Production and Use. *Encyclopedia of Life Support Systems* (EOLSS), Available from URL: http://www.eolss.net/ebooks/Sample%20Chapters/C08/E3-18-04-03.pdf

[30] Ghosh, D., Deb, A., Bera, S., Sengupta, R., Patra, K.K. (2007). Measurement of natural radioactivity in chemical fertilizer and agricultural soil: evidence of high alpha activity. *Environ Geochem Health,* DOI 10.1007/s10653-007-9114-0

[31] Glaser, B. (2007). Prehistorically modified soils of central Amazonia: a model for sustainable agriculture in the twenty-first century. *Philosophical Transactions of the Royal Society – Biological Sciences*, 362: 1478.

[32] Glaser, B., Lehmann, J., Zech,W. (2002). Ameliorating physical and chemical properties of highly weathered soils in the tropics with charcoal – a review, *Biol Fertil Soils,* 35:219–230.

[33] Glaser, B., Haumaier, L., Guggenberger and G., Zech, W. (2001). The 'Terra Preta' phenomenon: a model for sustainable agriculture in the humid tropics, *Naturwissenschaften* 88: 1.

[34] Glaser, B., Haumaier, L., Guggenberger, G., Zech, W., The 'Terra Preta' phenomenon: a model for sustainable agriculture in the humid tropics, *Naturwissenschaften,* Volume 88, Number 1 / February, 2001, pages 37-41.

[35] Gold, M.V. (1994). *Sustainable agriculture: definitions and terms.* SRB 94-05. Beltsville, MD: U.S. Department of Agriculture.

[36] Golley F, Baudry J, Berry RJ, Bornkamm R, Dahlberg K, Jansson A-M, King J, Lee J, Lenz R, Sharitz R et al. (1992). What is the road to sustainability? *INTECOL Bull* 20: 15–20.

[37] Grimm, K. PHOSPHORITES FEED PEOPLE: FINITE FERTLIZER ORES IMPACT CANADIAN AND GLOBAL FOOD SECURITY. Available from URL: http://www.eos.ubc.ca/personal/grimm/phosphorites.html

[38] Hamer, U., Marschner, B., Brodowski, S., Amelung, W., (2004). Interactive priming of black carbon and glucose mineralisation. *Organic Geochemistry,* 35, no. 7 (July): 823-830.

[39] Heerink, N., Kuyvenhoven, A. and Van Wijk, S.M. (2001). Economic policy reforms and sustainable land use in LDCs: issues and approaches. In N. Heerink, H. Van Keulen and M. Kuiper, eds. *Economic policy reforms and sustainable land use in LDCs. Recent advances in quantitative analysis*, pp. 1–20. Heidelberg, Germany, Physica Verlag.

[40] Heimlich, R. (2003). Agricultural Resources and Environmental Indicators. US Department of Agriculture, Agriculture Handbook No. (AH722).

[41] Helsel Z.R. (1992). Energy and alternatives for fertilizer and pesticide use. *Energy in Farm Production*, Volume 6 (ed. R.C. Fluck), pp. 177–201. New York: Elsevier.

[42] Hera, C. (1996). The role of inorganic fertilizers and their management practices. In: C. Rodriguez-Barrueco (Ed.), *Fertilizer and Environment*, Kluwer Academic Publishers, Dordrecht, 131-149.

[43] Hodges, S.C. SOIL FERTILITY BASICS. Soil Science Extension North Carolina State University, Available from URL: http://www.soil.ncsu.edu

[44] Houghton, J.T., Ding, Y., Griggs, D.J., Noguer, M., van der Linden, P.J., Dai, X., Johnson, C.A., and Maskell, K. (eds.)., IPCC (2001). *Climate Change 2001: The*

Scientific Basis, Intergovernmental Panel on Climate Change, Cambridge University Press. Cambridge, United Kingdom.

[45] Houghton, J.T., Meira Filho, L.G., Callander, B.A., Harris, N., Kattenberg, A. and Maskell,K. (eds.). *IPCC (1996) Climate Change 1995: The Science of Climate Change.* Intergovernmental Panel on Climate Change, Cambridge University Press. Cambridge, United Kingdom.

[46] Houghton, R.A. (2003). *Revised estimates of the annual net flux of carbon to the atmosphere from changes in land use and land management 1850–2000.* Tellus 55B: 378–390.

[47] Huettl, R.F., *Liming and fertilization as mitigation tools in declining forest ecosystems, Water, Air and Soil Pollution,* Volume 44, Numbers 1-2 / March, 1989, Pages: 98-118.

[48] Hüttl, R.F., Schneider, B.U., Forest ecosystem degradation and rehabilitation. *Ecological Engineering,* Volume 10, Issue 1, February 1998, Pages 19-31.

[49] Ingham,R.E, J. A. Trofymow, Elaine R. Ingham, and David C. Coleman Interactions of Bacteria, Fungi, and their Nematode Grazers: Effects on Nutrient Cycling and Plant Growth. *Ecological Monographs,* 1985 March.Volume 55, Issue 1.

[50] Johnston, H.W. 1954. The solubilization of "insoluble" phosphate. II – A quantitative and comparative study of the action of selected aliphatic acids on tricalcium phosphate. *N. Z. J. Sci. Tech. B.*, 36: 49–55.

[51] Kant, S., Kafkafi, U. Mitigation by Crop Management. Available from URL: http://www.plantstress.com/articles/min_deficiency_m/mitigation.htm

[52] Kauwenbergh, S. J. Van. *Natural and agricultural (fertilizer) inputs of Cadmium.* International Fertilizer Development Center.

[53] Khan, S.S, Mulvaney, R.L., Ellsworth, T.R., Boast, C.W., The Myth of Nitrogen Fertilization for Soil Carbon Sequestration, *J Environ Qual.,*2007; 36: 1821-1832.

[54] Kongshaug, G., Bockman, O.C., Kaarstad, O. and Morka, H. (1992). Inputs of trace elements to soils and plants. *Proc. Chemical Climatology and Geomedical Problems,* Norsk Hydro, Oslo, Norway.

[55] Kpomblekou, K. and Tabatabai, M.A. 1994. Effect of organic acids on release of phosphorus from phosphate rocks. *J Soil Sci.*, 158: 442–453.

[56] Lampkin, N. (1990). *Organic Farming. Published by Farming Press Books,* United Kingdom.

[57] Lehmann, J., Rondon, M. Bio-Char Soil Management on Highly Weathered Soils in the Humid Tropics. Available from URL: ttp://www.css.cornell.edu/faculty/lehmann/

[58] Lehmann, J., Gaunt, J. and Rondon, M. (2006). "Bio-char sequestration in terrestrial ecosystems –a review". *Mitigation and Adaptation Strategies for Global Change,* 11, 403–427.

[59] Lehmann, J., da Silva Jr, J.P., Rondon, M., Cravo, M.S., Greenwood, J., Nehls, T., Steiner, C. and Glaser, B. (2002). 'Slash-and-char – a feasible alternative for soil fertility management in the central Amazon?', *Proceedings of the 17th World Congress of Soil Science,* (pp. 1–12) Bangkok, Thailand.

[60] Lehmann, J., da Silva Jr, J.P., Steiner, C., Nehls, T., Zech, W. and Glaser, B. (2003). Nutrient availability and leaching in an archaeological Anthrosol and a Ferralsol of the Central Amazon basin: fertilizer, manure and charcoal amendments. *Plant and Soil,* 249: 343–357.

[61] Lewandrowski, J., Peters, M., Jones, C., House, R., Sperow, M., Eve, M., Paustian, K., (2004). Economics of Sequestering Carbon in the U.S. Agricultural Sector, US Department of Agriculture, Economic research service, Technical Bulletin Number 1909.

[62] Lickacz, J., Penny, D. Soil Organic Matter. *Agri Facts. Practical Information for Alberta's Agriculture Industry*, Agdex 536-1.

[63] Marschner, H. (1993). *Mineral nutrition of higher plants*. London, Academic Press Ltd., Harcourt Brace and Co. Publishers.

[64] Matson, P. A., Naylor, R. and Ortiz-Monasterio, I. (1998). Integration of environmental, agronomic, and economic aspects of fertilizer management. *Science*, 280: 112–115.

[65] McFarlin, R. F., Florida institute of phosphate Research (FIPR) (1995). Establishing Vegetation Cover on phosphogypsum in Florida. Publication number: 01-086-116.

[66] Mclaughlin, N.B., Hiba, A., Wall, G.J., King, D.J. Comparison of energy inputs for inorganic fertilizer and manure based corn production, *Canadian Agricultural Engineering,* Vol. 42, No. 1 January/February/March 2000.

[67] Menzel, R.G. (1968). Uranium, radium, and thorium content in phosphate rocks and their possible radiation hazard. *J. Agr. Food Chem*., 16, 231-284.

[68] Michinori, N. Microbial Fertilizers in Japan. Available from URL: http://www.agnet.org/library/eb/430/, accessed November 19, 2007

[69] Mizuta, K., Matsumoto, T., Hatate, Y., Nishihara, K. and Nakanishi T. (2004). Removal of nitratenitrogenfrom drinking water using bamboo powder charcoal, *Bioresource Technology* 95, 255–257.

[70] Miller, B.W. (2006).The long-term effects of a single phosphorus fertilizer application on phosphorus availability in forest soils., 18th World Congress of Soil Science July 9-15, 2006. Philadelphia, Pennsylvania, USA.

[71] Naylor, R. (1996). Energy and resource constraints on intensive agricultural production. *Annu. Rev. Energy Environ*., 21: 99–123.

[72] OECD (1993) *World Energy Outlook*. OECD, Paris.

[73] Oosterhuis, F.H., Brouwer, F.M., Wijnants, H.J. (2000). A possible EU wide charge on cadmium in phosphate fertilisers: Economic and environmental implications. Report number E-00/02, commissioned by the European Commission.

[74] Paustian, K., Antle, J. M., Sheehan, J., Paul, E.A., (2006). *Agriculture's role in greenhouse gas mitigation*. Prepared for the Pew Center on Global Climate Change.

[75] Pohlman, A.A. and McColl, G.J. 1986. Kinetics of metal dissolution from forest soils by organic acids.,*J. Env. Qual*., 15: 86–92.

[76] Radovic, L.R., Moreno-Castilla, C. and Rivera-Utrilla, J. (2001).Carbon materials as adsorbents inaqueous solutions', in L.R. Radovic (ed.). *Chemistry and Physics of Carbon,* (pp. 227–405) New York, Marcel Dekker.

[77] Ragothama KG (1999). Phosphate acquisition. *Annu Rev Plant Physiol Plant Mol Biol.,* 50: 665–693).

[78] Reicosky, D.C, Environmental Benefits of Soil Carbon Sequestration, Available from URL: http://www.depweb.state. pa.us/watersupply/lib/watersupply/ nnovTechForum-IIE-Reicosky.pdf

[79] Rehm, G. Schmitt, M., Lamb, J. Randall, G. Busman, L. 1998. *Understanding phosphorous fertilizers*. Communication and Educational Technology Services,

University of Minnesota Extension. Available from URL: Http://www.extension.umn.edu

[80] Romheld, V. (1998). The soil-root interface (rhizosphere): Its relationship to nutrient availability and plant nutrition. In: *International workshop on role of environmental and biological factors in acquisition of toxic and essential elements by plants*. Research Institute of Pomology and Floriculture, Skierniewice. pp. 41-58.

[81] Saito, M. and Marumoto,T. Inoculation with arbuscular mycorrhizal fungi: the status quo in Japan and the future prospects, *Plant and Soil*, IssueVolume 244, Numbers 1-2 / July, 2002 , 273-279.

[82] Sander, M., Pignatello, J. J. Characterization of charcoal adsorption sites for aromatic compounds: insights drawn from single-solute and bi-solute competitive experiments. *Environmental Science andamp; Technology,* 2005 (Vol. 39) (No. 6) 1606-1615

[83] SCENAR 2020 (2006). Scenario study on agriculture and the rural world. European Commission, Directorate-General Agriculture and Rural Development, Directorate G. Economic analysis and evaluation, G.4 Evaluation of measures applicable to agriculture (Contract No. 30 – CE – 0040087/00-08).

[84] Schactman DP, Reid RJ, Ayling SM (1998). Phosphorus uptake by plants: from soil to cell. *Plant Physiol* 116: 447–453.

[85] Schimel, D. S., Braswell, B. H., Holland, E., McKeown, R., Ojima, D. S., Painter, T. H., Parton, W. J. and Townsend, A. R. (1994). *Global Biogeochem. Cycles,* 8, 279–294.

[86] Shoemaker; R., McGranahan, D., McBride, W. (2006). Agriculture and Rural Communities Are Resilient to High Energy Costs. Rising energy prices may prompt farmers and rural residents to make tradeoffs in their production practices and daily lives. Economic Research Service/USDA, WW.ERS.USDA.GOV/AMBERWAVES

[87] Singh, C.P. and Amberger, A. (1990). Humic substances in straw compost with rock phosphate. Bio. *Wastes*, 31: 165–174.

[88] Skjemstad, J.O, Reicosky, D.C, Wilts, A.R, McGowan, J.A. (2002). Charcoal Carbon in U.S. Agricultural Soils, *Published in Soil Sci. Soc. Am. J.,* 66:1249–1255.

[89] Skjemstad, J.O., R.C. Dalal, L.J. Janik, and J.A. McGowan. (2001). Changes in chemical nature of soil organic carbon in Vertisols, under wheat in southeastern Queensland. *Aust. J. Soil Res*., 39:343–359.

[90] Smil, V. 2000. Phosphorus in the environment: natural flows and human interferences. *Annu. Rev. Energy Environ.*, 25: 53–88.

[91] Spaargaren, O. (2006). Mineral soils influenced by human activity: Anthrosols (new reference group: Technosols), In: 3rd European summer school on soil survey / Michéli, E., Panagos, P., Jones, A., Montanarella, L. Luxembourg: Office for Official Publications of the European Communities, (EUR 22193 EN) - p. 16-1 - 16-5.

[92] Steen, P. (1998). Phosphorus Availability in the 21st Century: Management of a Nonrenewable Resource. *Phosphorus and Potassium* 217. Available from URL: www.nhm.ac.uk/mineralogy/phos/pandk217/steen.htm

[93] Steiner, C. (2006). Slash and Char as Alternative to Slash and Burn - Soil charcoal amendments maintain soil fertility and establish a carbon sink. Dissertation, Faculty of Biology, Chemistry and Geosciences, University of Bayreuth, Germany.

[94] Tiessen, H, ed. (1995). Phosphorus in the Global Environment - Transfers, Cycles and Management, SCOPE/UNEP International Phosphorus Project, Published by John and Sons Ltd.

[95] Tiessen, H. ed. 1995. *Scope 54: Phosphorus in the Global Environment: Transfers, Cycles and Management.* Wiley, 1995. Available from URL: www. icsu-scope.org

[96] Tiessen, H. *Scope 54: Phosphorus in the Global Environment - Transfers, Cycles and Management*, 1995, 480 pp.

[97] Tilman, D. et al. (2001). Forecasting agriculturally driven global environmental change. *Science*, 292: 281–284.

[98] Tilman, D., Cassman, K. G., Matson, P.A., Naylor, R., and Polasky, S. 2002.Agricultural sustainability and intensive production practices. *Nature,* 418: 671-677.

[99] Tilman D, Fargione J, Wolff B, D'Antonio C, Dobson A, Howarth R, Schindler D, Schlesinger WH, Simberloff D, Swackhamer D (2001). Forecasting agriculturally driven global environmental change. *Science,* 292: 281–284.

[100] Trumbore S.E., Potential responses of soil organic carbon to global environmental change. *Proc. Natl. Acad. Sci.,* August 1997. Vol. 94, pp. 8284–8291,

[101] University of York (2007, October 21). Hungry Microbes Share Out The Carbon In The Roots Of Plants. *Science Daily*. Retrieved May 9, 2008, Available from URL: http://www.sciencedaily.com /releases/2007/10/071018123523.htm

[102] US Biochar Provisions S. (1884). Salazar Act (July 26, 2007).

[103] U.S. Department of Agriculture (USDA). (2003). Agricultural resources and environmental indicators, 2003. Agriculture Handbook no. 722 (AH222), Washington DC: Economic Research Service. Available from URL: www.ers.usda.gov/publications/arei/ah722/.

[104] US Department of Agriculture, U.S. Fertilizer Use and Price Data Sheets. Available from URL: http://www.ers.usda.gov/Data/FertilizerUse/

[105] U.S. EPA – Inventory of U.S. Greenhouse Gas Emissions and sinks: 1990 – 2006 April 15, 2008, U.S. Environmental Protection Agency, EPA 430-R-08-005.

[106] U.S. Epa Environmental protection Agency. About phosphogypsum. Available from URL: http://epa.gov/radiation/neshaps/subpartr/about.html#general

[107] von Uexku ll, H.R., Mutert, E. (1995) Global extent, development and economic impact of acid soils. *Plant Soil,* 171: 1–15.

[108] Vance, C. P., Symbiotic Nitrogen Fixation and Phosphorus Acquisition. Plant Nutrition in a World of Declining Renewable Resources, *Plant Physiology*, October 2001, Vol. 127, pp. 390–397.

[109] Vance CP, Graham PH, Allan DL (2000). Biological nitrogen fixation: phosphorus Ba critical future need? In FO Pederosa, M Hungria, MG Yates, WE Newton, eds, *Nitrogen Fixation from Molecules to Crop Productivity*. Kluwer Academic Publishers, Dordrecht, The Netherlands, pp 509–518.

[110] Vassilev, N., Vassileva, M. 2003. Biotechnological solubilization of rock phosphate on media containing agro-industrial wastes. *Applied Microbiology and Biotechnology*. Publisher Springer Berlin / Heidelberg ISSN0175-7598. Volume 61, Numbers 5-6.

[111] Vitousek, P. M., Mooney, H. A., Lubchenco, J. and Melillo, J. M. 1997. Human domination of earth's ecosystems. *Science*, 277: 494–499.

[112] Vitousek, P. M. et al.Human alteration of the global nitrogen cycle: sources and consequences. *Ecol. Applic.,* 7, 737–750 (1997).

[113] Wardle,D.A., Nilsson, M.C., Olle Zackrisson, Fire-Derived Charcoal Causes Loss of Forest Humus, *Science*, Vol 320 2 May 2008

[114] Watson, R.A. and Osborne, P.L. (1979). An algal pigment ratio as an indicator of the nitrogen supply to phytoplankton in three Norfolk broads. *Freshwater Biol.*, 9, 585-594.

[115] Wiebe, K., Gollehon, N., editors, Agricultural Resources and Environmental Indicators, 2006 Edition, United State Department of Agrculture, *Economic Iinformation Bulletin* 16.

[116] Woolf, D. (2008). Biochar as a soil amendment: A review of the environmental implications. Available from URL: http://www.orgprints.org/13268/01/ Biochar_as_a_soil_amendment_-_a_review.pdf

[117] Williams, S. Fertilizer: Application (Organic vs Inorganic.) Available from URL: http://gardenline.usask.ca/misc/fertili2.html

[118] Wright, S. Glomalin, A Manageable Soil Glue. Available from URL: www.ars.usda.gov/sp2UserFiles/Place/12650400/glomalin/brochure.pdf.

[119] Worrell, E., Phylipsen, D., Einstein, D., Martin, N., (2000) Energy Use and Energy Intensity of the U.S. Chemical Industry, Energy Analysis Department Environmental Energy Technologies Division Ernest Orlando Lawrence Berkeley National Laboratory, LBNL-44314, Available from URL: http://www.energystar.gov/ia/ business /industry/industrial_LBNL-44314.pdf

[120] Smith, P., D. Martino, Z. Cai, D. Gwary, H. Janzen, P. Kumar, B. McCarl, S. Ogle, F. O'Mara, C. Rice, B. Scholes, O. Sirotenko (2007): Agriculture. In Climate Change 2007: Mitigation. Contribution of Working Group III to the Fourth Assessment Report of the Intergovernmental Panel on Climate Change [B. Metz, O.R. Davidson, P.R. Bosch, R. Dave, L.A. Meyer (eds)], Cambridge University Press, Cambridge, United Kingdom and New York, NY, USA.Available from URL:http://www.mnp.nl/ipcc/ pages_media/FAR4docs/final_pdfs_ar4/Chapter08.pdf

[121] FAO, 2003: *World Agriculture: Towards 2015/2030. An FAO Perspective*. FAO, Rome, 97 pp.

[122] International Trade Centre (UNCTAD/WTO), *Research Institute of Organic Agriculture* (*FiBL*): *Organic farming and climate change, 2007.*

[123] C Leifert, E. Rembiałkowska, J.H. Nielson, J.M. Cooper, G. Butler, L. Lueck, Effects of organic and 'low input' production methods on food quality and safety, 3rd QLIF Congress, Hohenheim, Germany, March 20-23, 2007, Available from URL: http://orgprints.org/10482/01/Leifert-etal-2007-food-quality-safety.pdf

[124] NJF-Seminar 369, Organic farming for a new millennium, status and future challenge, Published by Nordic Association of Agricultural Scientists (NJF), Section I: Soil, Water and Environment Swedish University of Agricultural Sciences, Alnarp, Sweden June 15-17, 2005, ISSN 1653-2015.

[125] Nutritional quality of organic versus conventional fruits, vegetables, and grains, Worthington, Virginia, MS, ScD, CNS, *The Journal of Alternative and omplementary Medicine*, 7(2): 161-73, 1991. Available from RL:http://www.nutrition4health.org/nohanews/NNSp02NutQualOrganicVsConv.htm

[126] Vall, M.P, Vidal, C., Eurostat, Nitrogen in agriculture, available from URL: http://ec.europa.eu/agriculture/envir/report/en/nitro_en/report.htm#sche2

[127] Niggli, U., Fließbach, A. Low greenhouse gas agriculture, Mitigation and adaptation potential of sustainable farming system. FAO, 2008, available from URL: www.fao.org/organicag)

[128] FAO/WHO Codex Alimentarius Commission *Guidelines for the Production, Processing, Labelling and Marketing of Organically Produced Foods*

INDEX

A

B

C

D

E

F

G

H

I

J

K

L

M

N

O

P

Q

R

S

T

U

V

W

Y

Z